I0476619

NOAA Technical Report NESDIS 142-7

Regional Climate Trends and Scenarios for the U.S. National Climate Assessment

Part 7. Climate of Alaska

Washington, D.C.
January 2013

U.S. DEPARTMENT OF COMMERCE
National Oceanic and Atmospheric Administration
National Environmental Satellite, Data, and Information Service

NOAA TECHNICAL REPORTS

National Environmental Satellite, Data, and Information Service

The National Environmental Satellite, Data, and Information Service (NESDIS) manages the Nation's civil Earth-observing satellite systems, as well as global national data bases for meteorology, oceanography, geophysics, and solar-terrestrial sciences. From these sources, it develops and disseminates environmental data and information products critical to the protection of life and property, national defense, the national economy, energy development and distribution, global food supplies, and the development of natural resources.

Publication in the NOAA Technical Report series does not preclude later publication in scientific journals in expanded or modified form. The NESDIS series of NOAA Technical Reports is a continuation of the former NESS and EDIS series of NOAA Technical Reports and the NESC and EDS series of Environmental Science Services Administration (ESSA) Technical Reports.

Copies of earlier reports may be available by contacting NESDIS Chief of Staff, NOAA/ NESDIS, 1335 East-West Highway, SSMC1, Silver Spring, MD 20910, (301) 713-3578.

NOAA Technical Report NESDIS 142-7

Regional Climate Trends and Scenarios for the U.S. National Climate Assessment

Part 7. Climate of Alaska

Brooke C. Stewart, Kenneth E. Kunkel, Laura E. Stevens, and Liqiang Sun
Cooperative Institute for Climate and Satellites (CICS)
North Carolina State University and NOAA's National Climatic Data Center (NCDC)
Asheville, NC

John E. Walsh
University of Alaska Fairbanks
Fairbanks, AK
 and
University of Illinois at Urbana-Champaign
Champaign, IL

U.S. DEPARTMENT OF COMMERCE
Rebecca Blank, Acting Secretary

National Oceanic and Atmospheric Administration
Dr. Jane Lubchenco, Under Secretary of Commerce for Oceans and Atmosphere
and NOAA Administrator

National Environmental Satellite, Data, and Information Service
Mary Kicza, Assistant Administrator

PREFACE

This document is one of series of regional climate descriptions designed to provide input that can be used in the development of the National Climate Assessment (NCA). As part of a sustained assessment approach, it is intended that these documents will be updated as new and well-vetted model results are available and as new climate scenario needs become clear. It is also hoped that these documents (and associated data and resources) are of direct benefit to decision makers and communities seeking to use this information in developing adaptation plans.

There are nine reports in this series, one each for eight regions defined by the NCA, and one for the contiguous U.S. The eight NCA regions are the Northeast, Southeast, Midwest, Great Plains, Northwest, Southwest, Alaska, and Hawai'i/Pacific Islands.

These documents include a description of the observed historical climate conditions for each region and a set of climate scenarios as plausible futures – these components are described in more detail below.

While the datasets and simulations in these regional climate documents are not, by themselves, new, (they have been previously published in various sources), these documents represent a more complete and targeted synthesis of historical and plausible future climate conditions around the specific regions of the NCA.

There are two components of these descriptions. One component is a description of the historical climate conditions in the region. The other component is a description of the climate conditions associated with two future pathways of greenhouse gas emissions.

Historical Climate

The description of the historical climate conditions was based on an analysis of core climate data (the data sources are available and described in each document). However, to help understand, prioritize, and describe the importance and significance of different climate conditions, additional input was derived from climate experts in each region, some of whom are authors on these reports. In particular, input was sought from the NOAA Regional Climate Centers and from the American Association of State Climatologists. The historical climate conditions are meant to provide a perspective on what has been happening in each region and what types of extreme events have historically been noteworthy, to provide a context for assessment of future impacts.

Future Scenarios

The future climate scenarios are intended to provide an internally consistent set of climate conditions that can serve as inputs to analyses of potential impacts of climate change. The scenarios are not intended as projections as there are no established probabilities for their future realization. They simply represent an internally consistent climate picture using certain assumptions about the future pathway of greenhouse gas emissions. By "consistent" we mean that the relationships among different climate variables and the spatial patterns of these variables are derived directly from the same set of climate model simulations and are therefore physically plausible.

These future climate scenarios are based on well-established sources of information. No new climate model simulations or downscaled data sets were produced for use in these regional climate reports.

The use of the climate scenario information should take into account the following considerations:

1. All of the maps of climate variables contain information related to statistical significance of changes and model agreement. This information is crucial to appropriate application of the information. Three types of conditions are illustrated in these maps:

   a. The first condition is where most or all of the models simulate statistically significant changes and agree on the direction (whether increasing or decreasing) of the change. If this condition is present, then analyses of future impacts and vulnerabilities can more confidently incorporate this direction of change. It should be noted that the models may still produce a significant range of magnitude associated with the change, so the manner of incorporating these results into decision models will still depend to a large degree on the risk tolerance of the impacted system.

   b. The second condition is where the most or all of the models simulate changes that are too small to be statistically significant. If this condition is present, then assessment of impacts should be conducted on the basis that the future conditions could represent a small change from present or could be similar to current conditions and that the normal year-to-year fluctuations in climate dominate over any underlying long-term changes.

   c. The third condition is where most or all of the models simulate statistically significant changes but do not agree on the direction of the change, i.e. a sizeable fraction of the models simulate increases while another sizeable fraction simulate decreases. If this condition is present, there is little basis for a definitive assessment of impacts, and, separate assessments of potential impacts under an increasing scenario and under a decreasing scenario would be most prudent.

2. The range of conditions produced in climate model simulations is quite large. Several figures and tables provide quantification for this range. Impacts assessments should consider not only the mean changes, but also the range of these changes.

3. Several graphics compare historical observed mean temperature and total precipitation with model simulations for the same historical period. These should be examined since they provide one basis for assessing confidence in the model simulated future changes in climate.

   a. Temperature Changes: Magnitude. In most regions, the model simulations of the past century simulate the magnitude of change in temperature from observations; the southeast region being an exception where the lack of century-scale observed warming is not simulated in any model.

   b. Temperature Changes: Rate. The *rate* of warming over the last 40 years is well simulated in all regions.

   c. Precipitation Changes: Magnitude. Model simulations of precipitation generally simulate the overall observed trend but the observed decade-to-decade variations are greater than the model observations.

In general, for impacts assessments, this information suggests that the model simulations of temperature conditions for these scenarios are likely reliable, but users of precipitation simulations may want to consider the likelihood of decadal-scale variations larger than simulated by the models. It should also be noted that accompanying these documents will be a web-based resource with downloadable graphics, metadata about each, and more information and links to the datasets and overall descriptions of the process.

1. INTRODUCTION

The Global Change Research Act of 1990[1] mandated that national assessments of climate change be prepared not less frequently than every four years. The last national assessment was published in 2009 (Karl et al. 2009). To meet the requirements of the act, the Third National Climate Assessment (NCA) report is now being prepared. The National Climate Assessment Development and Advisory Committee (NCADAC), a federal advisory committee established in the spring of 2011, will produce the report. The NCADAC Scenarios Working Group (SWG) developed a set of specifications with regard to scenarios to provide a uniform framework for the chapter authors of the NCA report.

This climate document was prepared to provide a resource for authors of the Third National Climate Assessment report, pertinent to the state of Alaska. The specifications of the NCADAC SWG, along with anticipated needs for historical information, guided the choices of information included in this description of Alaskan climate. While guided by these specifications, the material herein is solely the responsibility of the authors and usage of this material is at the discretion of the NCA report authors.

This document has two main sections: one on historical conditions and trends, and the other on future conditions as simulated by climate models. The historical section concentrates on temperature and precipitation, primarily based on analyses of data from the National Weather Service's (NWS) Cooperative Observer Network, which has been in operation since the late 19th century. Additional climate features are discussed based on the availability of information. The future simulations section is exclusively focused on temperature and precipitation.

With regard to the future, the NCADAC, at its May 20, 2011 meeting, decided that scenarios should be prepared to provide an overall context for assessment of impacts, adaptation, and mitigation, and to coordinate any additional modeling used in synthesizing or analyzing the literature. Scenario information for climate, sea-level change, changes in other environmental factors (such as land cover), and changes in socioeconomic conditions (such as population growth and migration) have been prepared. This document provides an overall description of the climate information.

In order to complete this document in time for use by the NCA report authors, it was necessary to restrict its scope in the following ways. Firstly, this document does not include a comprehensive description of all climate aspects of relevance and interest to a national assessment. We restricted our discussion to climate conditions for which data were readily available. Secondly, the choice of climate model simulations was also restricted to readily available sources. Lastly, the document does not provide a comprehensive analysis of climate model performance for historical climate conditions, although a few selected analyses are included.

The NCADAC directed the "use of simulations forced by the A2 emissions scenario as the primary basis for the high climate future and by the B1 emissions scenario as the primary basis for the low climate future for the 2013 report" for climate scenarios. These emissions scenarios were generated by the Intergovernmental Panel on Climate Change (IPCC) and are described in the IPCC Special Report on Emissions Scenarios (SRES) (IPCC 2000). These scenarios were selected because they incorporate much of the range of potential future human impacts on the climate system and because

---

[1] http://thomas.loc.gov/cgi-bin/bdquery/z?d101:SN00169:|TOM:/bss/d101query.html

there is a large body of literature that uses climate and other scenarios based on them to evaluate potential impacts and adaptation options. These scenarios represent different narrative storylines about possible future social, economic, technological, and demographic developments. These SRES scenarios have internally consistent relationships that were used to describe future pathways of greenhouse gas emissions. The A2 scenario "describes a very heterogeneous world. The underlying theme is self-reliance and preservation of local identities. Fertility patterns across regions converge very slowly, which results in continuously increasing global population. Economic development is primarily regionally oriented and per capita economic growth and technological change are more fragmented and slower than in the other storylines" (IPCC 2000). The B1 scenario describes "a convergent world with…global population that peaks in mid-century and declines thereafter…but with rapid changes in economic structures toward a service and information economy, with reductions in material intensity, and the introduction of clean and resource-efficient technologies. The emphasis is on global solutions to economic, social, and environmental sustainability, including improved equity, but without additional climate initiatives" (IPCC 2000).

The temporal changes of emissions under these two scenarios are illustrated in Fig. 1 (left panel). Emissions under the A2 scenario continually rise during the 21st century from about 40 gigatons (Gt) CO_2-equivalent per year in the year 2000 to about 140 Gt CO_2-equivalent per year by 2100. By contrast, under the B1 scenario, emissions rise from about 40 Gt CO_2-equivalent per year in the year 2000 to a maximum of slightly more than 50 Gt CO_2-equivalent per year by mid-century, then falling to less than 30 Gt CO_2-equivalent per year by 2100. Under both scenarios, CO_2 concentrations rise throughout the 21st century. However, under the A2 scenario, there is an acceleration in concentration trends, and by 2100 the estimated concentration is above 800 ppm. Under the B1 scenario, the rate of increase gradually slows and concentrations level off at about 500 ppm by 2100. An increase of 1 ppm is equivalent to about 8 Gt of CO_2. The increase in concentration is considerably smaller than the rate of emissions because a sizeable fraction of the emitted CO_2 is absorbed by the oceans.

The projected CO_2 concentrations are used to estimate the effects on the earth's radiative energy budget, and this is the key forcing input used in global climate model simulations of the future. These simulations provide the primary source of information about how the future climate could evolve in response to the changing composition of the earth's atmosphere. A large number of modeling groups performed simulations of the 21st century in support of the IPCC's Fourth Assessment Report (AR4), using these two scenarios. The associated changes in global mean temperature by the year 2100 (relative to the average temperature during the late 20th century) are about +6.5°F (3.6°C) under the A2 scenario and +3.2°F (1.8°C) under the B1 scenario with considerable variations among models (Fig. 1, right panel).

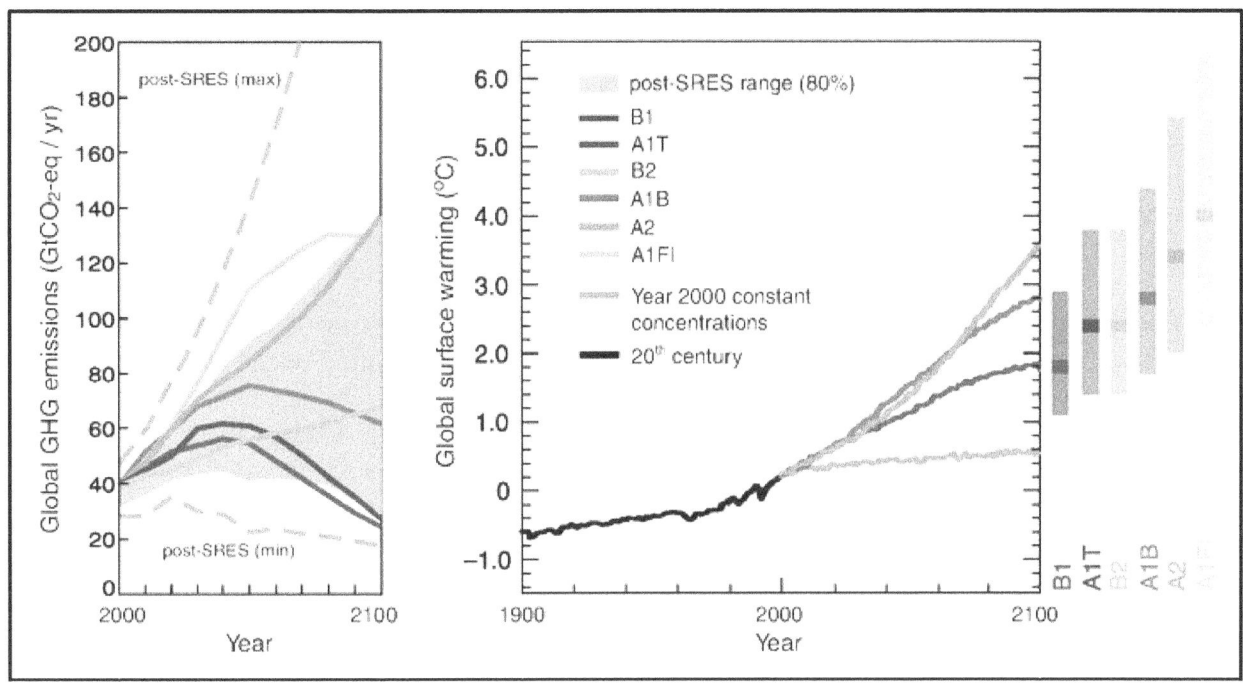

Figure 1. Left Panel: Global GHG emissions (in GtCO₂-eq) in the absence of climate policies: six illustrative SRES marker scenarios (colored lines) and the 80ᵗʰ percentile range of recent scenarios published since SRES (post-SRES) (gray shaded area). Dashed lines show the full range of post-SRES scenarios. The emissions include CO₂, CH₄, N₂O and F-gases. Right Panel: Solid lines are multi-model global averages of surface warming for scenarios A2, A1B and B1, shown as continuations of the 20ᵗʰ-century simulations. These projections also take into account emissions of short-lived GHGs and aerosols. The pink line is not a scenario, but is for Atmosphere-Ocean General Circulation Model (AOGCM) simulations where atmospheric concentrations are held constant at year 2000 values. The bars at the right of the figure indicate the best estimate (solid line within each bar) and the likely range assessed for the six SRES marker scenarios at 2090-2099. All temperatures are relative to the period 1980-1999. From IPCC AR4, Sections 3.1 and 3.2, Figures 3.1 and 3.2, IPCC (2007b).

In addition to the direct output of the global climate model simulations, the NCADAC approved "the use of both statistically- and dynamically-downscaled data sets". "Downscaling" refers to the process of producing higher-resolution simulations of climate from the low-resolution outputs of the global models. The motivation for use of these types of data sets is the spatial resolution of global climate models. While the spatial resolution of available global climate model simulations varies widely, many models have resolutions in the range of 100-200 km (~60-120 miles). Such scales are very large compared to local and regional features important to many applications. For example, at these scales mountain ranges are not resolved sufficiently to provide a reasonably accurate representation of the sharp gradients in temperature, precipitation, and wind that typically exist in these areas.

Statistical downscaling achieves higher-resolution simulations through the development of statistical relationships between large-scale atmospheric features that are well-resolved by global models and the local climate conditions that are not well-resolved. The statistical relationships are developed by comparing observed local climate data with model simulations of the recent historical climate. These relationships are then applied to the simulations of the future to obtain local high-resolution projections. Statistical downscaling approaches are relatively economical from a computational perspective, and thus they can be easily applied to many global climate model simulations. One underlying assumption is that the relationships between large-scale features and local climate conditions in the present climate will not change in the future (Wilby and Wigley 1997). Careful consideration must also be given when deciding how to choose the appropriate predictors because statistical downscaling is extremely sensitive to the choice of predictors (Norton et al. 2011).

Dynamical downscaling is much more computationally intensive but avoids assumptions about constant relationships between present and future. Dynamical downscaling uses a climate model, similar in most respects to the global climate models. However, the climate model is run at a much higher resolution but only for a small region of the earth (such as North America) and is termed a "regional climate model (RCM)". A global climate model simulation is needed to provide the boundary conditions (e.g., temperature, wind, pressure, and humidity) on the lateral boundaries of the region. Typically, the spatial resolution of an RCM is 3 or more times higher than the global model used to provide the boundary conditions. With this higher resolution, topographic features and smaller-scale weather phenomena are better represented. The major downside of dynamical downscaling is that a simulation for a region can take as much computer time as a global climate model simulation for the entire globe. As a result, the availability of such simulations is limited, both in terms of global models used for boundary conditions and time periods of the simulations (Hayhoe 2010).

Section 3 of this document (Future Regional Climate Scenarios) responds to the NCADAC directives by incorporating analyses from multiple sources. The core source is the set of global climate model simulations performed for the IPCC AR4, also referred to as the Climate Model Intercomparison Project phase 3 (CMIP3) suite. These have undergone extensive evaluation and analysis by many research groups. A second source is a set of statistically-downscaled data sets based on the CMIP3 simulations. A third source is a set of dynamically-downscaled simulations, driven by CMIP3 models. A new set of global climate model simulations is being generated for the IPCC Fifth Assessment Report (AR5). This new set of simulations is referred to as the Climate Model Intercomparison Project phase 5 (CMIP5). These scenarios do not incorporate any CMIP5 simulations as relatively few were available at the time the data analyses were initiated.
As noted earlier, the information included in this document is primarily concentrated around analyses of temperature and precipitation. This is explicitly the case for the future scenarios sections; due in large part to the short time frame and limited resources, we capitalized on the work of other groups on future climate simulations, and these groups have devoted a greater effort to the analysis of temperature and precipitation than other surface climate variables.

Climate models have generally exhibited a high level of ability to simulate the large-scale circulation patterns of the atmosphere. These include the seasonal progression of the position of the jet stream and associated storm tracks, the overall patterns of temperature and precipitation, the occasional occurrence of droughts and extreme temperature events, and the influence of geography on climatic patterns. There are also important processes that are less successfully simulated by models, as noted by the following selected examples.

Climate model simulation of clouds is problematic. Probably the greatest uncertainty in model simulations arises from clouds and their interactions with radiative energy fluxes (Dufresne and Bony 2008). Uncertainties related to clouds are largely responsible for the substantial range of global temperature change in response to specified greenhouse gas forcing (Randall et al. 2007). Climate model simulation of precipitation shows considerable sensitivities to cloud parameterization schemes (Arakawa 2004). Cloud parameterizations remain inadequate in current GCMs. Consequently, climate models have large biases in simulating precipitation, particularly in the tropics. Models typically simulate too much light precipitation and too little heavy precipitation in both the tropics and middle latitudes, creating potential biases when studying extreme events (Bader et al. 2008).

Climate models also have biases in simulation of some important climate modes of variability. The El Niño-Southern Oscillation (ENSO) is a prominent example. In some parts of the U.S., El Niño and La Niña events make important contributions to year-to-year variations in conditions. Climate models have difficulty capturing the correct phase locking between the annual cycle and ENSO (AchutaRao and Sperber 2002). Some climate models also fail to represent the spatial and temporal structure of the El Niño - La Niña asymmetry (Monahan and Dai 2004). Climate simulations over the U.S. are affected adversely by these deficiencies in ENSO simulations.

The model biases listed above add additional layers of uncertainty to the information presented herein and should be kept in mind when using the climate information in this document.

The representation of the results of the suite of climate model simulations has been a subject of active discussion in the scientific literature. In many recent assessments, including AR4, the results of climate model simulations have been shown as multi-model mean maps (e.g., Figs. 10.8 and 10.9 in Meehl et al. 2007). Such maps give equal weight to all models, which is thought to better represent the present-day climate than any single model (Overland et al. 2011). However, models do not represent the current climate with equal fidelity. Knutti (2010) raises several issues about the multi-model mean approach. These include: (a) some model parameterizations may be tuned to observations, which reduces the spread of the results and may lead to underestimation of the true uncertainty; (b) many models share code and expertise and thus are not independent, leading to a reduction in the true number of independent simulations of the future climate; (c) all models have some processes that are not accurately simulated, and thus a greater number of models does not necessarily lead to a better projection of the future; and (d) there is no consensus on how to define a metric of model fidelity, and this is likely to depend on the application. Despite these issues, there is no clear superior alternative to the multi-model mean map presentation for general use. Tebaldi et al. (2011) propose a method for incorporating information about model variability and consensus. This method is adopted here where data availability make it possible. In this method, multi-model mean values at a grid point are put into one of three categories: (1) models agree on the statistical significance of changes and the sign of the changes; (2) models agree that the changes are not statistically significant; and (3) models agree that the changes are statistically significant but disagree on the sign of the changes. The details on specifying the categories are included in Section 3.

2. REGIONAL CLIMATE TRENDS AND IMPORTANT CLIMATE FACTORS

2.1. Description of Data Sources

The observational analyses are based primarily on a set of 26 Alaskan surface weather observing stations with hourly data[2]. Station criteria were based on period of record (POR) and percent completeness of those records. The general minimum requirements for inclusion of stations were that they cover the time period from 1949 to 2011 and that those records be at least 90% complete, though most have 95% or greater coverage. Percent completeness of data was calculated individually for each station, by parameter and by decade. Exceptions regarding criteria for inclusion were made for several stations. Anchorage was included despite its approximate 75% coverage in the 1950s because it is a major city with excellent records for the duration of its POR. Bettles, having approximately 85% coverage in the 1950s, was included due to the need for stations located within the interior basin of Alaska. Big Delta and Northway have a few decades with data availability just below 90%, but were also included (along with Tanana, with approximately 80% coverage in the 1970s) due to their interior locations. Other stations that were included with some decades with data availability below 90% are Sitka, Iliamna, and Ketchikan. These stations were also included in order to improve spatial coverage. The parameters used were daily maximum air temperature, daily minimum air temperature, and 24-hour precipitation totals.

2.2. General Description of Alaskan Climate

Alaska's climate is influenced by four main factors: latitude, altitude, continentality (proximity to the ocean versus the continental interior), and the seasonal distribution of sea ice (ACRC 2012). The state's vast expanse and geographical variation lend to a variety of climate types and should be examined regionally.

The first region, in which maritime influences on climate are strong, consists of the Southeast, South Coast, and Southwestern Islands as well as west-central Alaska during the warm season. These areas experience a maritime climate with high precipitation and moderate temperatures. The state's highest annual average temperatures (45-47 °F) and highest precipitation amounts (100-200 inches) are found in this region (Shulski and Wendler 2007; WRCC 2011).

Seasonal distribution of sea ice plays a major role in the climate of west central Alaska, which experiences a maritime influence until the open water of the Bering Sea is cutoff by sea ice, leaving the west coast with a more continental-like climate. Sea ice is generally established along the coast by late fall and remains until late spring (Shulski and Wendler 2007; WRCC 2011).

A transitional zone between maritime and continental climates exists in portions of Bristol Bay, Cook Inlet, and the southern portion of the Copper River Basin. The area is largely cutoff from

---

[2] Stations and regions include Anchorage (south central), Annette (southeast), Barrow (Arctic), Bethel (west coast), Bettles (interior), Big Delta (interior), Cold Bay (west coast), Cordova (southeast), Fairbanks (interior), Gulkana (south central), Homer (south central), Iliamna (west coast), Juneau (southeast), Kenai (south central), Ketchikan (southeast), King Salmon (west coast), Kodiak (south central), Kotzebue (west coast), McGrath (interior), Nome (west coast), Northway (interior), St. Paul (west coast), Sitka (southeast), Talkeetna (south central), Tanana (interior), Yakutat (southeast).

maritime influence by mountains, but experiences moderate temperatures in comparison to a continental climate (Shulski and Wendler 2007; WRCC 2011).

The second region is Interior Alaska, bounded by the Brooks and Alaska Ranges, experiences a truly continental climate, with large temperature variability, low humidity, and relatively light and irregular precipitation. Summers are warm and sunny, while winters are long and cold, with frequent low-level temperature inversions caused by radiational cooling at the surface (Shulski and Wendler 2007; WRCC 2011).

The third region is the area north of the Brooks Range, also known as Alaska's Arctic region, where the state's lowest annual average temperatures are found. Bordered on the north by the Arctic Sea, coastal areas of this region are affected by the moderating effect of open waters. Summers are cool and cloudy along the coast, and temperatures remain too low to allow any tree growth. Farther inland, climate conditions are more continental with warmer summers and cooler winters. Precipitation is relatively light, but commonly under-reported, as frequent high winds result in gauge undercatch; blizzard conditions are not uncommon in the winter (Shulski and Wendler 2007).

Average annual temperatures (Fig. 2) across Alaska vary widely. Southern coastal areas, heavily influenced by ocean waters, are the only areas with annual averages above freezing, while the state's high-latitude location and high elevations keep much of the state below freezing. The large north-south temperature gradient is due mainly to latitudinal variations in insolation and the maritime Pacific influence. Incoming solar radiation is not only a function of the hours of sunlight per day, but also of the solar elevation angle, which remains quite low throughout the year. Areas with a continental climate experience the largest variations in temperature on an interannual and interseasonal basis. Interior Alaska, for example, sees nearly a 90°F range in temperature between summer highs and winter lows. By contrast, southern areas under a maritime influence have an interseasonal range of only 30-40°F (Shulski and Wendler 2007).

Average annual precipitation amounts also vary greatly across Alaska, with a nearly exponential decrease in totals from south to north (Fig. 3). Coastal mountain ranges in the southeastern panhandle may receive as much as 200 inches per year, while totals drop to 60 inches south of the Alaska Range, 12 inches in the Interior, and less than 6 inches in the Arctic on average (WRCC 2011). These large variations are due to differences in elevation, temperature, moisture availability, and topography. Storm patterns have a significant impact on seasonal timing of precipitation. Most of the state sees precipitation maxima in the summer when storms track eastward and gain moisture from ice-free waters; or when daytime surface heating leads to convective precipitation. By contrast, southern regions see precipitation maxima in late fall to early winter under the influence of the Aleutian Low (Shulski and Wendler 2007).

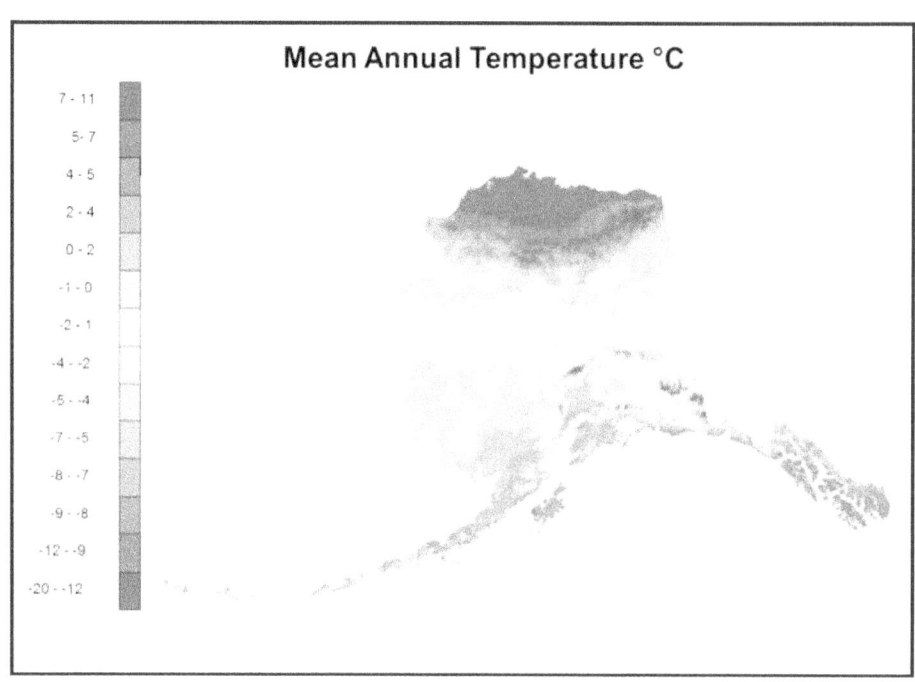

Figure 2. Mean annual temperature (°C) in Alaska for 2000-2009. Maps were produced by the Scenarios Network for Alaska Planning (SNAP) at the University of Alaska Fairbanks using data from the Climatic Research Unit (CRU) at the University of East Anglia, downscaled using the base climatology produced by the PRISM Climate Group at Oregon State University.

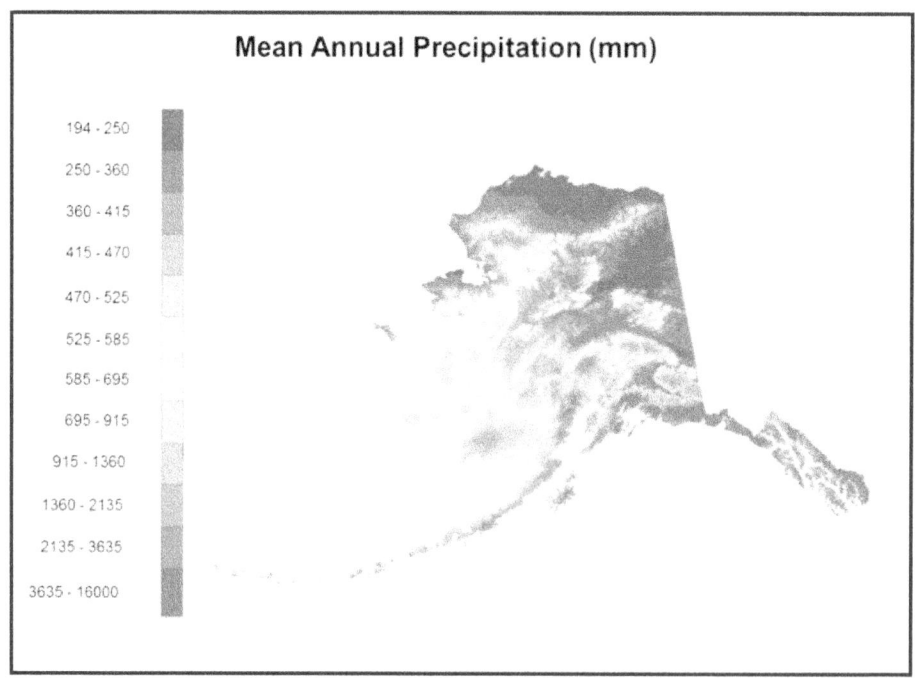

Figure 3. Mean annual precipitation (rain and melted snowfall, mm) in Alaska for 2000-2009. Maps were produced by the Scenarios Network for Alaska Planning (SNAP) at the University of Alaska Fairbanks using data from the Climatic Research Unit (CRU) at the University of East Anglia, downscaled using the base climatology produced by the PRISM Climate Group at Oregon State University.

2.3. Important Climate Factors

Alaska experiences a wide range of extreme weather and climate events that affect ecosystems and human society, including infrastructure. This discussion is meant to provide general information about these types of weather and climate phenomena. These include:

2.3.1. Rising Temperatures

Warming temperatures pose serious threats to Alaska, where average temperatures are so near freezing; a small change in temperature can result in large impacts. For example, forests in Alaska support two major economic veins for the state: tourism and timber. Tourism depends partially on the presence of healthy, beautiful forests while the lumber industry relies on live, healthy trees. Increasing temperatures (more specifically, a decrease in occurrence of extreme cold temperatures) have resulted in increased over-winter survival of bark beetles, and a consequent increase in the number of acres of forest destroyed by these insects. Dead trees combined with warmer, drier conditions leave the forests more vulnerable to wildfires (Karl et al. 2008). Another consequence of the recent warming is the reduction of the length of the snow season in Alaska (Liston and Hiemstra 2011), with impacts on winter recreation and the timing of the spring break-up. The term "break-up" period refers to the interval between melt onset and the final disappearance of snow and ice from non-glaciated areas. It corresponds with the major discharge pulse in streams and rivers carrying meltwater to the ocean. The "break-up" generally marks the end of the winter recreation season. Finally, the retreat of Alaskan glaciers represents a threat to freshwater supplies and tourism. In the offshore region, warming ocean temperatures have had serious impacts on other economic resources, such as the contamination of oyster beds by parasites and bacteria.

2.3.2. Thawing Permafrost

Much of Alaska is underlain by permafrost, or perennially frozen ground. Permafrost is characterized by an "active" layer that thaws from the surface downward (to maximum depths of 0.3-1.5 meters) during the warm season, only to refreeze during autumn (Callaghan et al. 2011). Active layer depths range from several tens of centimeters on the Alaskan North Slope, to as much as several meters in areas of marginally stable permafrost. Below the active layer, permafrost depths range from several meters to several hundred meters. Permafrost depths and temperatures are determined primarily by two climate variables: mean annual temperature and wintertime snow depth, as well as by soil properties and the slope (aspect) of the ground. Because snow effectively insulates the ground during winter, the boundary of permafrost is typically near the -1° or -2°C isotherm of mean annual temperature. However, soil properties and slope variations result in large areas of discontinuous permafrost. At present, much of the Alaskan Interior (between the Brooks and Alaska Ranges) is a zone of discontinuous permafrost, while the area north of the Brooks Range is essentially continuous permafrost (Smith et al. 2010).

Thawing permafrost causes major issues for various types of infrastructure. Roads, runways, and buildings may shift, break, or collapse as the ground beneath them becomes soft and sinks. Crude oil, a major mineral export for the state, is transported from Prudhoe Bay via the trans-Alaska pipeline. The pipeline required the use of expensive and innovative technology to disperse heat away from the permafrost that lies beneath it in hopes of extending the structure's lifetime (Karl et al. 2009).

2.3.3. Coastal Storms

Coastal storms increasingly threaten Alaska's coasts with rapid erosion; a result of the longer ice-free seasons (loss of protective barrier) and associated wave action, as well as degraded permafrost (from warmer average temperatures). Coastal erosion is happening at accelerating rates (tens of feet per year) in some areas of Alaska, causing entire communities to consider mass relocation. Storms also interfere with shipping operations, commercial fishing fleets, and other marine traffic (Karl et al. 2009).

2.3.4. Sea Ice Decline

Warming waters pose significant threats to Alaska's commercial fishing industry as marine ecosystems undergo alterations. In addition to providing much of the United States' commercial fishing catch, many native communities are dependent upon marine species for their food supply. Sea ice decline causes problematic changes in the marine ecosystem food chain, resulting in drastic changes to marine species composition and population (Karl et al. 2009). Loss of sea ice also affects several marine species (e.g., seals, walruses, and polar bears) that depend on its use as hunting and/or breeding grounds.

2.3.5. Floods

Floods come from a variety of unique sources in Alaska and can have disastrous impacts on local communities. Increasing snowpacks or rapid springtime temperature increases can cause unseasonable and excessive glacial and snow melt at higher elevations, resulting in flooding. When coastal areas lack a protective sea-ice barrier, storm surge from high-wind events can result in flooding. High-latitude regions of the globe are susceptible to floods caused by the ice jams on rivers. In addition to upstream flooding caused by the damming effect of an ice jam, the dislodging of an ice jam can release large quantities of backed-up water and huge chunks of ice downstream into local communities, having catastrophic results (Shulski and Wendler 2007).

2.3.6. Drought

Drought also has regionally unique impacts in Alaska. Low river flow from drought conditions can hinder mobility for rural areas where rivers are used as a major mode of transportation. Drought also leaves the vast boreal forests of Alaska particularly vulnerable to wildfire (Shulski and Wendler 2007; see also Kasischke et al. 2010).

2.4. Climatic Trends

2.4.1. Temperature

Long-term temperature trends are clearly non-linear (Fig. 4). Annual departures from the long-term mean show mostly negative (cool) anomalies from 1949 to 1976, then mostly positive (warm) anomalies for the last three decades. It is important to note that the observed stepwise shift from cool to warm anomalies around 1976 corresponds to a phase shift in the Pacific Decadal Oscillation (PDO) from negative to positive. In the positive phase of the PDO, there is increased southerly flow and warm air advection into Alaska during the winter, resulting in mostly positive temperature anomalies. When examined regionally and seasonally, it is apparent that most of the warming has occurred in winter and spring for all regions and that the Interior region has experienced the greatest overall warming (Table 1). Aside from the rapid warming seen in 1976, little additional warming has occurred during this positive phase of the PDO (see Fig. 4; Shulski and Wendler 2007; ACRC 2012).

Table 1. Total change in mean seasonal and annual temperature for 19 first-order surface weather observing stations from 1949-2011. Table from Alaska Climate Research Center (ACRC 2012).

Total Change in Mean Seasonal and Annual Temperature (°F), 1949 - 2011

Region	Location	Winter	Spring	Summer	Autumn	Annual
Arctic	Barrow	7.3	4.8	3.2	4.5	4.9
Interior	Bettles	7.2	4.7	1.8	1.6	3.9
	Fairbanks	7.0	3.9	2.3	0.1	3.3
	Big Delta	9.0	3.6	1.1	0.2	3.4
	McGrath	7.4	4.7	2.5	1.0	3.9
West Coast	Kotzebue	6.6	1.8	2.7	1.7	3.2
	Nome	4.5	3.1	2.3	0.4	2.5
	Bethel	6.7	4.3	1.8	-0.1	3.2
	King Salmon	8.0	4.1	1.2	0.6	3.4
	St Paul	0.7	1.4	2.3	1.1	1.4
	Cold Bay	1.5	1.2	1.5	0.8	1.2
Southcentral	Talkeetna	8.9	5.4	2.9	2.5	4.9
	Gulkana	7.5	2.4	0.8	0.1	2.7
	Anchorage	5.8	3.5	1.4	1.7	3.2
	Homer	5.4	3.4	2.8	1.4	3.4
	Kodiak	1.3	2.0	1.1	-0.5	0.9
Southeast	Yakutat	5.4	3.0	2.1	0.7	2.8
	Juneau	6.4	2.9	2.0	1.3	3.1
	Annette	3.9	2.3	1.7	0.2	1.9
	Average	5.8	3.3	2.0	1.0	3.0

Alaska Climate Research Center   Geophysical Institute, University of Alaska Fairbanks

15

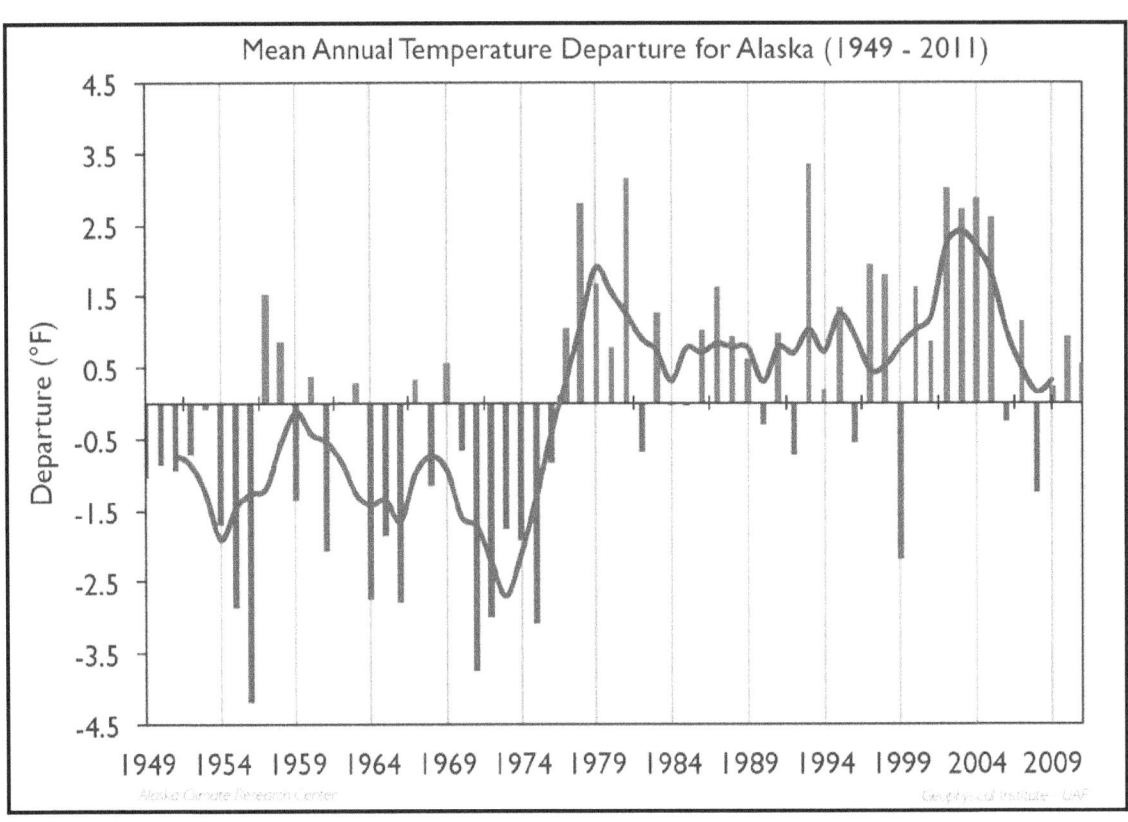

Figure 4. Annual temperature departure from the long-term mean (1949-2011) for 19 first-order surface weather observing stations in Alaska. Black line represents the 5-year running average. Figure from ACRC (2012).

Associated with the seasonal temperature changes in Table 1 are changes in the lengths of the frost-free season and the growing season. While comprehensive analyses of historical trends of these variables are lacking, evaluations have been performed for individual stations having long records. For Fairbanks, for example, the frost-free season length increased by 2 to 3 weeks since 1950, with most of the increase occurring in the past 25 years. The first date of running mean temperatures exceeding 55°F shows an advance of about 10 days since 1950, although the early part of the century (1905-1945) shows an opposite trend. While the vegetative leaf-out date has also advanced during the past three to four decades, the trend is not significant (R. Thoman, personal communication, Feb 14, 2011).

2.4.2. Precipitation

Average annual precipitation, when examined as a percent of the long-term average (Shulski and Wendler (2007); 1949-2005 is used as the averaging period for the long-term average in their analysis), shows a statewide increase of about 10%, with both regional and seasonal variation. Notable overall decreases in both seasonal and annual precipitation were observed at Annette (southeastern panhandle) and Barrow (Arctic coast). Long-term trends show nearly average precipitation from 1949 to 1965, then about 15 years of below average totals. Recent decades have seen precipitation amounts largely above average in Alaska (Shulski and Wendler 2007).

2.4.3. Extreme Heat and Cold

Extreme temperatures display similar regional and seasonal variation to those of mean temperatures in Alaska. In a study (Stewart 2011) examining 26 Alaskan observing stations from 1950 to 2008, the greatest increase in frequency of warm extremes (warmest 1% of daily high temperatures) and the greatest decrease in frequency of cold extremes (coldest 1% of daily lows) is found in spring. Nearly all stations exhibit this behavior (Table 2). The observed decrease in frequency of extreme cold events in winter has been more pronounced in the past few decades. Winter shows the next greatest increase in frequency of warm extremes and decrease in frequency cold extremes. Results become much more variable in summer and fall, with fewer stations observing this pattern in extreme temperatures (Stewart 2011).

Table 2. Percent of stations in each region displaying increasing trends in occurrence of warm extremes and decreasing trends in frequency of cold extremes in spring from 1950-2008 (Stewart 2011).

Region	Warm Extremes	Cold Extremes
Arctic*	100%	100%
West Central	100%	100%
Interior	100%	100%
Southwest	100%	100%
South Central	100%	100%
Southeast	71%	86%

*Only one station (Barrow) had sufficient data for this analysis.

Figure 5 shows values of cold wave and heat wave indices for the state of Alaska based on data obtained from the same 26 stations as for Table 2 and includes data through 2010. The indices are based on both the number of heat/cold waves. Cold waves defined as 4-day periods or 7-day periods that are colder than the threshold for a 1-in-5 year recurrence, and heat waves defined as 4-day periods or 7-day periods that are warmer than the threshold for a 1-in-5-year recurrence. Individual station events were first identified by sorting 4-day or 7-day running average mean temperature values and selecting the top Y/R events, where Y is the number of years with data and R is the recurrence interval (5 in this case). The total number of events for all stations in a given year is normalized by the number of stations with sufficient data for that year. These indices are representative of the frequency of the extreme hot/cold events. Coincident with a known phase shift in the Pacific Decadal Oscillation (PDO) from its negative (cold) phase to positive (warm) phase in 1976 (ACRC 2012), there is a clear decrease in the occurrence of 4-day and 7-day cold wave events. In Interior Alaska, for example, the number of days with temperatures -40°C (F) or colder at Fairbanks averaged between 10 and 20 days per year during the 1960s, but fewer than 5 per year during the past decade. While not as readily apparent as the decrease in occurrence of cold waves, the average heat wave index after 1976 is three times that of the average value seen prior to 1976.

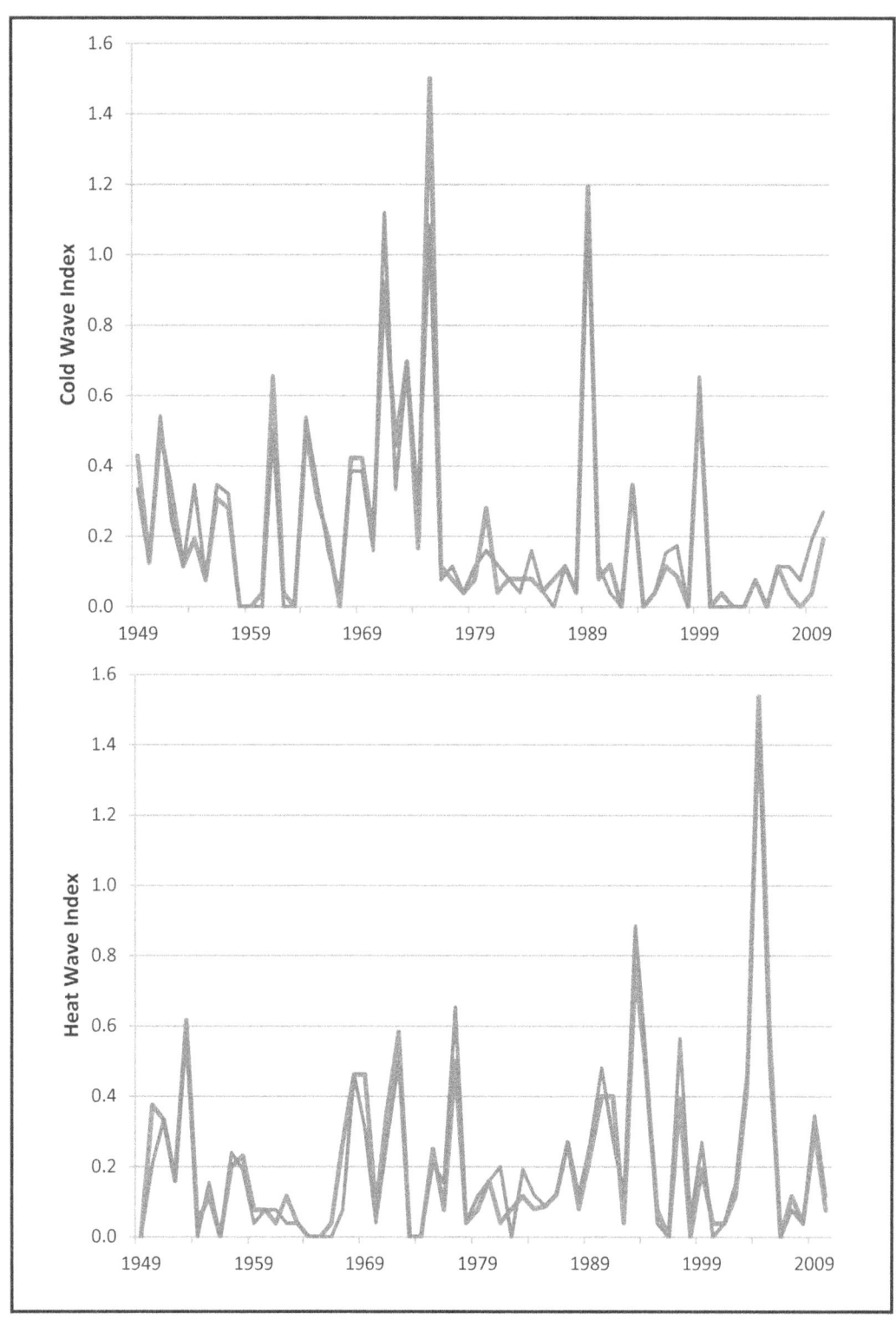

Figure 5. Time series (1949-2010) of an index for the occurrence of: cold waves defined as 4-day periods (blue) and 7-day periods (red) that are colder than the threshold for a 1-in-5 year recurrence (top), and heat waves defined as 4-day periods (blue) and 7-day periods (red) that are warmer than the threshold for a 1-in-5-year recurrence (bottom), for Alaska. Based on data from the National Climatic Data Center First-Order Surface Weather Observing Stations and based on methods from Kunkel et al. (1999).

2.4.4. Extreme Precipitation

As with average precipitation, occurrence of extreme precipitation events is highly variable and both regionally and seasonally dependent. Here, extreme precipitation events are defined as the heaviest 1% of three-day precipitation totals for each calendar season. Observations from 26 stations across Alaska were analyzed. (Three-day totals were used in order to capture events spanning consecutive days in the database of daily precipitation amounts). The most significant increases in extreme precipitation events are observed in the Southeast and West Central regions in spring, while the Arctic region shows statistically significant decreases in occurrence of extreme precipitation in all seasons except fall (Table 3). Results for summer and fall were largely insignificant with the exception of the Southeast in fall, which yielded high significant results for all but one station. With the exception of the Arctic, all regions have seen an increase in occurrence of extreme precipitation events in summer in the last few decades, though the statistical significance of this has not been tested (Stewart 2011).

Table 3. Percent of stations in each region displaying increasing trends in occurrence of extreme 3-day precipitation events from 1950 to 2008 (Stewart 2011).

Region	Winter	Spring	Summer	Fall
Arctic*	0%	0%	0%	0%
West Central	67%	100%	0%	67%
Interior	33%	50%	50%	50%
Southwest	50%	25%	25%	75%
South Central	40%	40%	40%	60%
Southeast	71%	57%	43%	57%

*Only one station (Barrow) had sufficient data for this analysis.

2.4.5. Storms

Coastal erosion rates in the Arctic are influenced by the frequency and intensity of coastal storms, by the duration of a protective buffer of sea ice and, ultimately, by changes of sea level. The relative roles played by changes in storminess and sea ice are a primary issue in the recent acceleration of coastal erosion. In a synthesis of Alaskan storm information based on the National Centers for Environmental Prediction (NCEP) storm track database (1953-2010)[3], no significant change in the overall frequency of strong storm events is detectable. However, along Alaska's northern and northwestern coasts, a significant increase in the number of strong storms has been observed when a protective sea ice cover is not present during the summer and autumn months. The increase of such events is primarily a function of the longer ice-free season. Overall storm frequencies show little trend south of the Bering Strait, where the loss of sea ice has been less (Perovich and Richter-Menge 2009; see subsequent section on Sea Ice).

---

[3] See http://www.cpc.ncep.noaa.gov/products/precip/CWlink/stormtracks/strack_alaska.shtml#outlook

2.4.6. Fires

During the decade of the 2000s, an average of 1,890,000 acres per year were burned in the interior sections of Alaska (17% of the landscape), which is 50% higher than in any previous decade since the 1940s (Kasischke et al. 2010). The increase in fire severity has occurred during a period of warmer spring seasons (Table 1) and lengthening growing seasons. Warmer spring temperature, earlier snow melt, and longer growing seasons increase the likelihood of a randomly occurring dry spell, often associated with a blocking ridge that persists longer than 10 days, at some point during the snow-free season. It is also thought that deeper active layers in permafrost areas allow fires to persist in the organic horizons of black spruce forests (Kasischke et al. 2010). The increase in fire occurrence has coincided with, and likely has been at least partially driven by, increases in lightning frequency since the 1990s (Faruch et al. 2011). However, trends in lightning frequency in Alaska have considerable uncertainty because there have been enhancements of the lightning detection network in Alaska over the past 15-20 years, introducing heterogeneities into the corresponding time series. The past several years have also seen unprecedented fire occurrence on the tundra of northern and western Alaska (Hu et al. 2010). Tundra fires have been associated with sea ice loss because both are favored by anomalous southerly advection of warm dry air into northern Alaska.

2.4.7. Sea Ice

Recent climate variability in the Arctic has resulted in significant changes of sea ice in the waters offshore of Alaska. These changes are most pronounced in the summer and autumn, which have been characterized by a striking retreat of sea ice (Fig. 6). Ice extent at the end of each of the past six summers (2007-2012) has been lower than at any time prior to 2007 in the satellite-based sea ice record (1979-2012). The most extreme minimum of all occurred in September 2012, when the minimum ice coverage (3.41 million km^2) was about 0.7 million km^2 less than in 2007, the year of the previous record minimum. The greatest change from previous decades has occurred in the Pacific sector of the Arctic. It appears that increased heat inflow through the Bering Strait from the North Pacific is one factor contributing to the extreme retreat of sea ice in the past decade (Shimada et al. 2006). Other contributing factors are thought to be the pattern of wind-forcing, the increase of air temperatures, and feedbacks associated with decreased albedo (Perovich and Richter-Menge 2009). On a pan-Arctic basis, the ice coverage at the time of the September minimum is now about 50% less than in the 1980s, and the ice is younger and thinner than at any time in the period of satellite coverage.

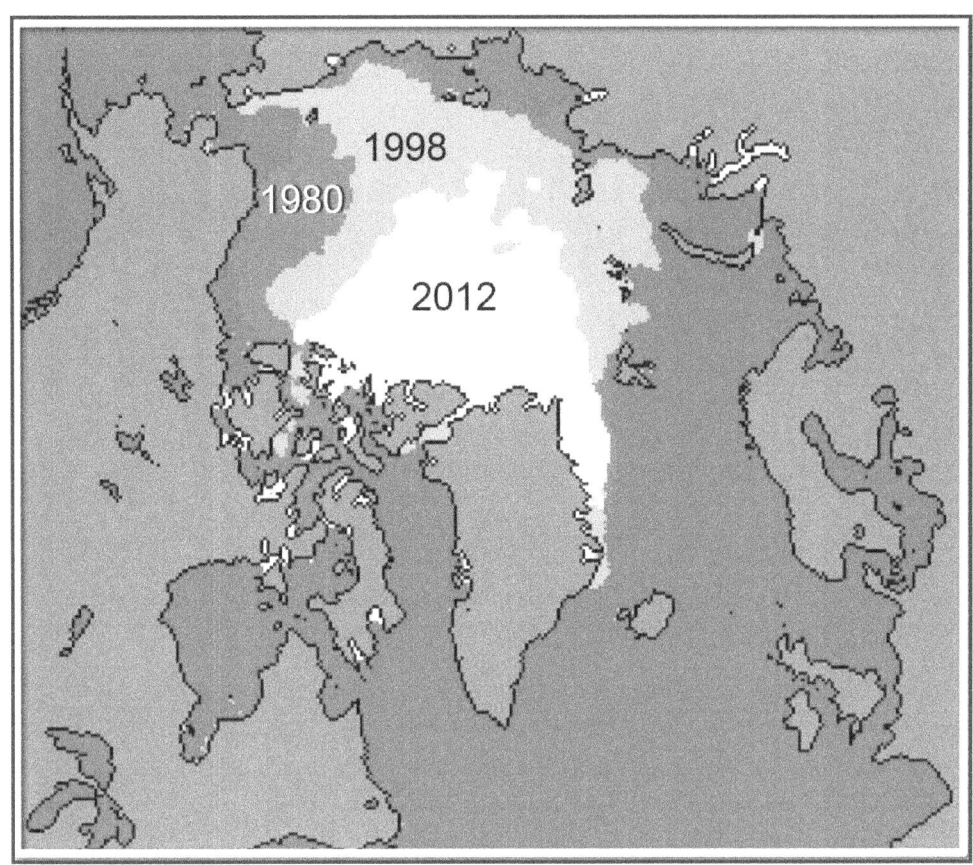

Figure 6. Extent of sea ice in September of 1980 (outer red boundary), 1998 (outer pink boundary) and 2012 (outer white boundary). Modified from Perovich and Richter-Menge (2009).

Despite the dramatic reduction in summer ice coverage since the 1980s, the reduction of wintertime ice extent has been considerably smaller. The wintertime (December-March) trends are typically several percent per decade, in contrast with the trend of more than 10% per decade during the summer season, with a net result of an increase in the area of seasonal (first-year) sea ice. The Bering Sea provides an example of the absence of a large trend of wintertime ice. As shown in Fig. 7, the Bering Sea ice cover that affects the western and southwestern Alaskan coasts has been characterized by large interannual to multi-year variability over the past several decades. Three recent winters, 2008-2010, had extensive ice cover by historical standards. While the winter of 2011 saw a return to more normal water temperatures and ice coverage, the Bering Sea ice cover of March 2012 was actually a new maximum record for the post-1979 satellite period. The implication of Fig. 7 is that sea ice - albeit seasonal and thin relative to the sea ice of the central Arctic - continues to be a player in the climatic regime of the western coast of Alaska during winter. This trend towards extensive areas of first-year ice is consistent with climate model projections of much greater loss in summer than in winter. However, the seasonal winter ice is relatively thin, more easily deformable, and more vulnerable to rapid melt in the spring, as has been characteristic of the past six years. The thinner ice is also less stable, increasing risks for near-shore activities such as subsistence hunting (AMAP 2011). Even in the central Arctic, the ice is much younger and thinner than in past decades (Stroeve et al. 2011).

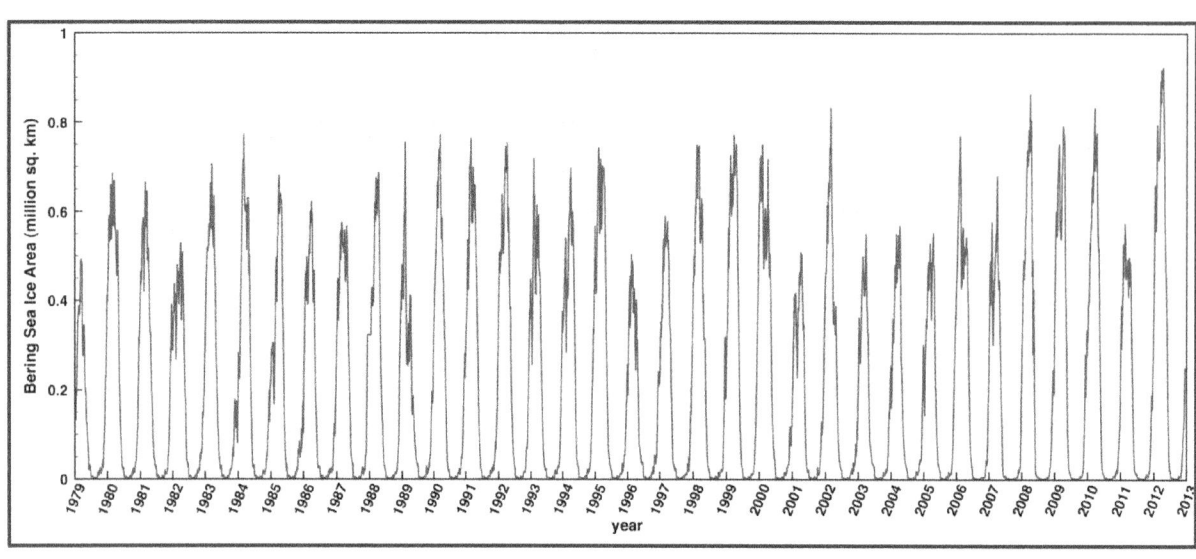

Figure 7. Time series of ice-covered area in the Bering Sea January 1 1979 through December 16 2012. Figure provided by William Chapman, University of Illinois. Data available from UIUC (2012).

3. FUTURE REGIONAL CLIMATE SCENARIOS

As noted above, the physical climate framework for the 2013 NCA report is based on climate model simulations of the future using the high (A2) and low (B1) SRES emissions scenarios. The resulting climate conditions are to be viewed as scenarios, not forecasts, and there are no explicit or implicit assumptions about the probability of occurrence of either scenario.

3.1. Description of Data Sources

This summary of future regional climate scenarios is based on the following model data sets:

- **Coupled Model Intercomparison Project phase 3 (CMIP3)** – Fifteen coupled Atmosphere-Ocean General Circulation Models (AOGCMs) from the World Climate Research Programme (WCRP) CMIP3 multi-model dataset (PCMDI 2012), as identified in the 2009 NCA report (Karl et al. 2009), were used (see Table 4). The spatial resolution of the great majority of these model simulations was 2-3° (a grid point spacing of approximately 100-200 miles), with a few slightly greater or smaller. All model data were re-gridded to a common resolution before processing (see below). The simulations from all of these models include:

  a) Simulations of the 20th century using best estimates of the temporal variations in external forcing factors (such as greenhouse gas concentrations, solar output, volcanic aerosol concentrations); and

  b) Simulations of the 21st century assuming changing greenhouse gas concentrations following both the A2 and B1 emissions scenarios. One of the fifteen models did not have a B1 simulation.

  These model simulations also serve as the basis for the following downscaled data set.

- **Downscaled CMIP3** – These temperature, precipitation, and growing season length data are downscaled products based on a subset of five CMIP3 models (Table 4) that were found to best simulate the seasonal cycles of temperature, precipitation, and sea-level pressure over Alaska during the final few decades of the 20th century in 20C3M simulations performed for AR4 (Walsh et al. 2008). These data are at 2 km resolution and were downscaled using Parameter-elevation Regressions on Independent Slopes Model (PRISM) data with a "delta" method applied to the CMIP3 model output. The downscaling has been carried out by the Scenarios Network for Alaska Planning (SNAP) program at the University of Alaska[4].

3.2. Analyses

Analyses are provided for the periods of 2021-2050, 2041-2070, and 2070-2099, with changes calculated with respect to the historical climate reference period of 1971-1999. These future periods will sometimes be denoted in the text by their mid-points of 2035, 2055, and 2085, respectively. The choice of 2021-2050 as the first time slice was based on the anticipated dominance of natural variability in near-term changes (i.e., several years to a decade), as externally forced signals become more prominent in changes as the timeframe lengthens.

---

[4] See http://www.snap.uaf.edu/

Table 4. Listing of the 15 models used for the CMIP3 simulations (left column). The 5 models used in the downscaled analyses are indicated (right column).

CMIP3 Models	Downscaled CMIP3
CCSM3	
CGCM3.1 (T47)	X
CNRM-CM3	
CSIRO-Mk3.0	
ECHAM5/MPI-OM	X
ECHO-G	
GFDL-CM2.0	
GFDL-CM2.1	X
INM-CM3.0	
IPSL-CM4	
MIROC3.2 (medres)	X
MRI-CGCM2.3.2	
PCM	
UKMO-HadCM3	X
UKMO-HadGEM1[5]	

Three different types of analyses are represented, described as follows:

- **Multi-model mean maps** – Model simulations of future climate conditions typically exhibit considerable model-to-model variability. In most cases, the future climate scenario information is presented as multi-model mean maps. To produce these, each model's data is first re-gridded to a common grid of approximately 2.8° latitude (~190 miles) by 2.8° longitude (~60-110 miles). Then, each grid point value is calculated as the mean of all available model values at that grid point. Finally, the mean grid point values are mapped. This type of analysis weights all models equally. Although an equal weighting does not incorporate known differences among models in their fidelity in reproducing various climatic conditions, a number of research studies have found that the multi-model mean with equal weighting is superior to any single model in reproducing the present-day climate (Overland et al. 2011). In most cases, the multi-model mean maps include information about the variability of the model simulations. In addition, there are several graphs that show the variability of individual model results. These should be examined to gain an awareness of the magnitude of the uncertainties in each scenario's future values.

---

[5] Simulations from this model are for the A2 scenario only.

- **Spatially-averaged products** – To produce these, all the grid point values within the Alaska region boundaries are averaged and represented as a single value. This is useful for general comparisons of different models, periods, and data sources. Because of the spatial aggregation, this product may not be suitable for many types of impacts analyses.

- **Probability density functions (pdfs)** – These are used here to illustrate the differences among models. To produce these, spatially-averaged values are calculated for each model simulation. Then, the distribution of these spatially-averaged values is displayed. This product provides an estimate of the uncertainty of future changes in a tabular form. As noted above, this information should be used as a complement to the multi-model mean maps.

3.3. Mean Temperature

While individual models show biases of Arctic surface air temperatures over pan-Arctic and smaller regional Arctic areas, the biases of the composited CMIP3 model simulations of Arctic air temperatures are generally quite small (less than 1° to 2 °C), although the average surface temperatures over the Arctic Ocean are too cold by about 2°C (Walsh et al. 2008). Surface-based temperature inversions are a special challenge in the Arctic, especially during the cold season. Most CMIP3 models over-estimate the inversion strength in the Arctic, although the nature of the biases differs between inversions over different regions: stable inversions over the Arctic Ocean/surrounding continents and unstable inversions over the North Atlantic (Medeiros et al. 2011).

Figure 8 shows the spatial distribution of multi-model mean simulated differences in average annual temperature for the three future time periods (2035, 2055, 2085) relative to the model reference period of 1971-1999, for both emissions scenarios, for the 14 (B1) or 15 (A2) CMIP3 models. The statistical significance regarding the change in temperature between each future time period and the model reference period was determined using a 2-sample t-test assuming unequal variances for those two samples. For each period (present and future climate), the mean and standard deviation were calculated using the 29 or 30 annual values. These were then used to calculate t. In order to assess the agreement between models, the following three categories were determined for each grid point, similar to that described in Tebaldi et al. (2011):

- *Category 1*: If less than 50% of the models indicate a statistically significant change then the multi-model mean is shown in color. Model results are in general agreement that simulated changes are within historical variations;
- *Category 2*: If more than 50% of the models indicate a statistically significant change, and less than 67% of the significant models agree on the sign of the change, then the grid points are masked out, indicating that the models are in disagreement about the direction of change;
- *Category 3*: If more than 50% of the models indicate a statistically significant change, and more than 67% of the significant models agree on the sign of the change, then the multi-model mean is shown in color with hatching. Model results are in agreement that simulated changes are statistically significant and in a particular direction.

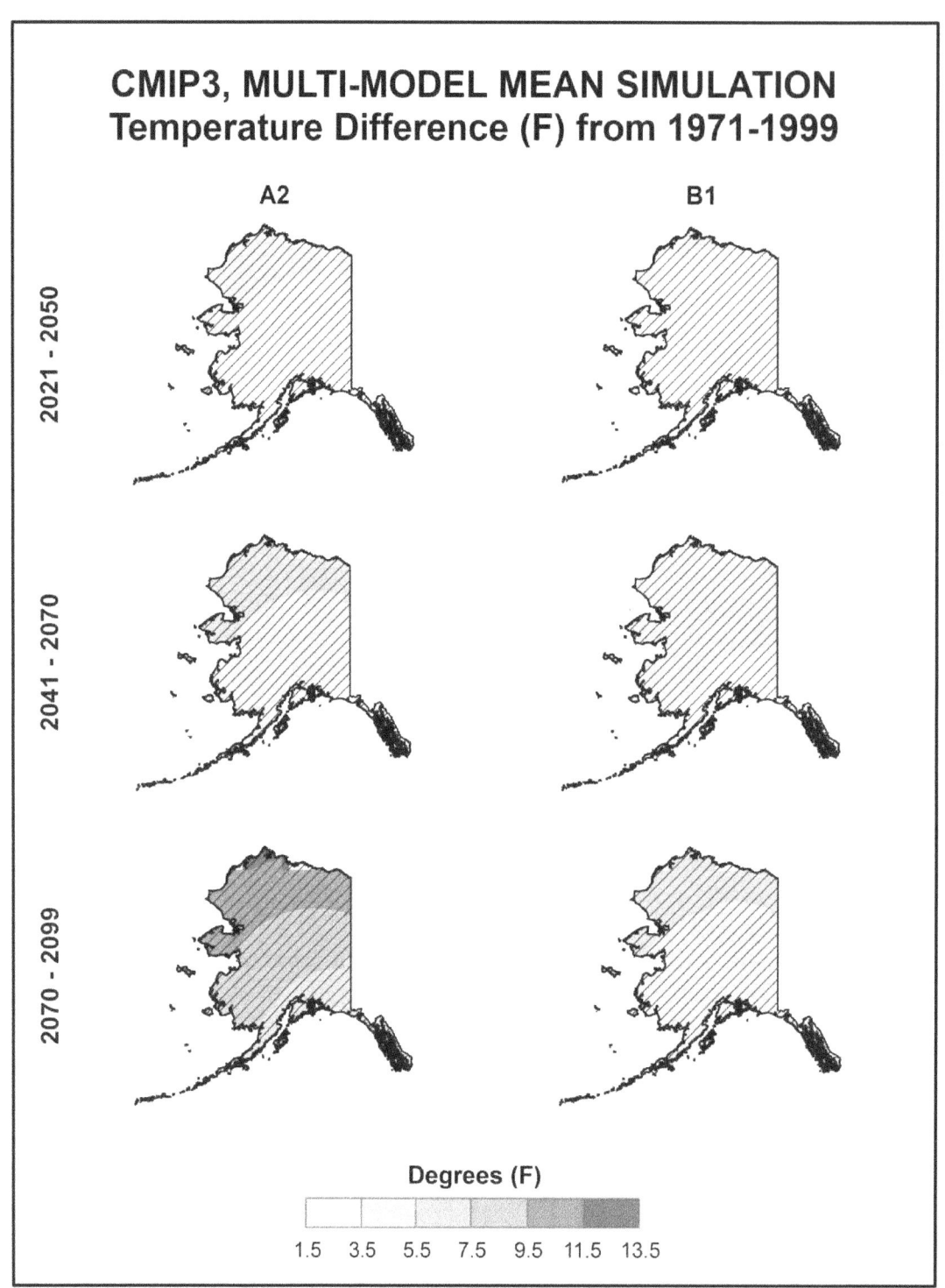

Figure 8 Simulated difference in annual mean temperature (°F) for Alaska, for each future time period (2021-2050, 2041-2070, and 2070-2099) with respect to the reference period of 1971-1999. These are multi-model means for the high (A2) and low (B1) emissions scenarios from the 14 (B1) or 15 (A2) CMIP3 global climate simulations. Color with hatching indicates that more than 50% of the models show a statistically significant change in temperature, and more than 67% agree on the sign of the change (see text). The far northwestern portion of Alaska sees the greatest amount of warming for both emissions scenarios and all time periods.

It can be seen from Fig. 8 that all three periods indicate an increase in temperature compared to the corresponding multi-model means for 1971-1999. The Inside Passage region shows the least amount of warming, with the greatest temperature changes being seen in the far northwest of Alaska. Warming increases over time, and the differences between the high (A2) and low (B1) emissions scenarios also increase over time.

Spatial variations are relatively small, especially for the B1 scenario. For 2035, both A2 and B1 values range between 1.5 and 5.5°. For 2055, warming for A2 is greater, ranging from 3.5 to 7.5°F. Increases by 2085 are larger still, with a 3.5-7.5°F range for B1 and a 7.5-13.5°F range for A2. The CMIP3 models indicate that temperature changes across Alaska, for all three future time periods and both emissions scenarios, are statistically significant. The models also agree on the sign of change, with all grid points satisfying category 3 above, i.e. the models are in agreement on temperature increases throughout the region for each future time period and scenario.

Figure 9 shows the simulated change in annual mean temperature for each future time period with respect to 1971-1999, for both emissions scenarios, averaged over the entire Alaska region for the 14 (B1) or 15 (A2) CMIP3 models. Both the multi-model mean and individual model values are shown. Temperature changes increase over time for both scenarios, with greater changes for A2. By 2085, simulated mean increases for the low (B1) emissions scenario are 5.0°F, and for A2 are greater, at 8.7°F.

A key overall feature is that the simulated temperature changes are similar in value for the high (A2) and low (B1) emissions scenarios for 2035, but largely different for 2085. This indicates that early in the 21st century, the multi-model mean temperature changes are relatively insensitive to the emissions path, whereas late 21st century changes are quite sensitive to the emissions pathway. This arises because atmospheric CO_2 concentrations resulting from the two different emissions scenarios do not considerably diverge from one another until around 2050 (see Fig. 1). It can also be seen from Fig. 9 that the range of individual model changes is quite large, with considerable overlap between the A2 and B1 results, even for 2085.

Figure 10 shows the simulated change in seasonal mean temperature for each future time period for the high (A2) emissions scenario, averaged over the entire Alaska region for the 15 CMIP3 models. Again, both the multi-model mean and individual model values are shown. For all seasons, warming is simulated to increase with time. Winter is simulated to see the greatest temperature increases, ranging from around 4°F in 2035 to more than 12°F in 2085. The spread of individual model values is large for all seasons, and also increases with time.

The distribution of changes in annual mean temperature for each future time period with respect to 1971-1999 for both emissions scenarios across the 14 (B1) or 15 (A2) CMIP3 models is shown in Table 5. These changes range from 1.8°F in 2035 for the low (B1) emissions scenario to 14.3°F in 2085 for the high (A2) emissions scenario. The inter-model range of temperature changes (i.e., the difference between the highest and lowest model values) is seen to increase for each future time period. The interquartile range (the difference between the 75th and 25th percentiles) varies between 0.6 to 1.6°F across the three time periods.

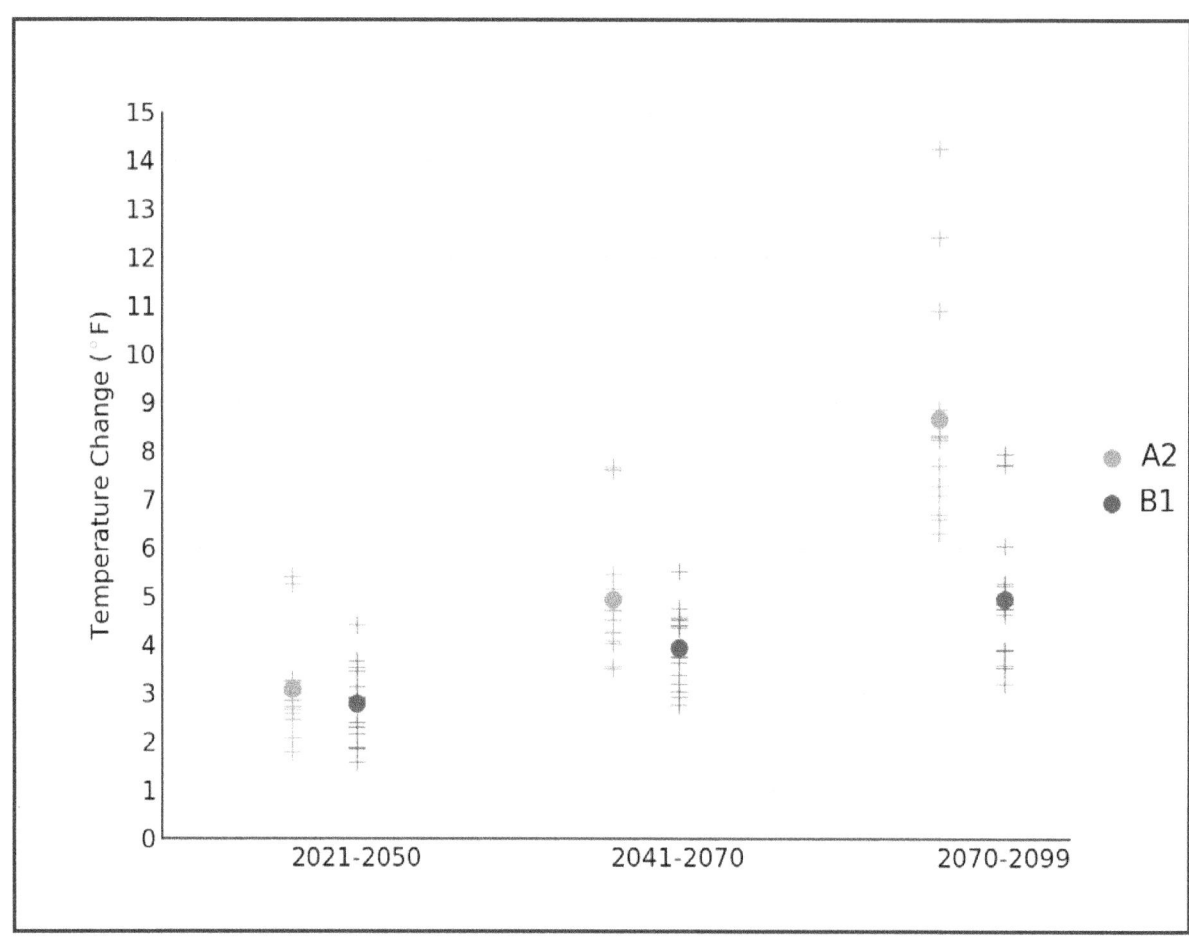

Figure 9. Simulated annual mean temperature change (°F) for Alaska, for each future time period (2021-2050, 2041-2070, and 2070-2099) with respect to the reference period of 1971-1999. Values are given for the high (A2) and low (B1) emissions scenarios for the 14 (B1) or 15 (A2) CMIP3 models. The small plus signs indicate each individual model and the circles depict the multi-model means. The range of model-simulated changes is large compared to the mean differences between A2 and B1 in the early and middle 21st century. By the end of the 21st century, the difference between A2 and B1 is comparable to the range of B1 simulations.

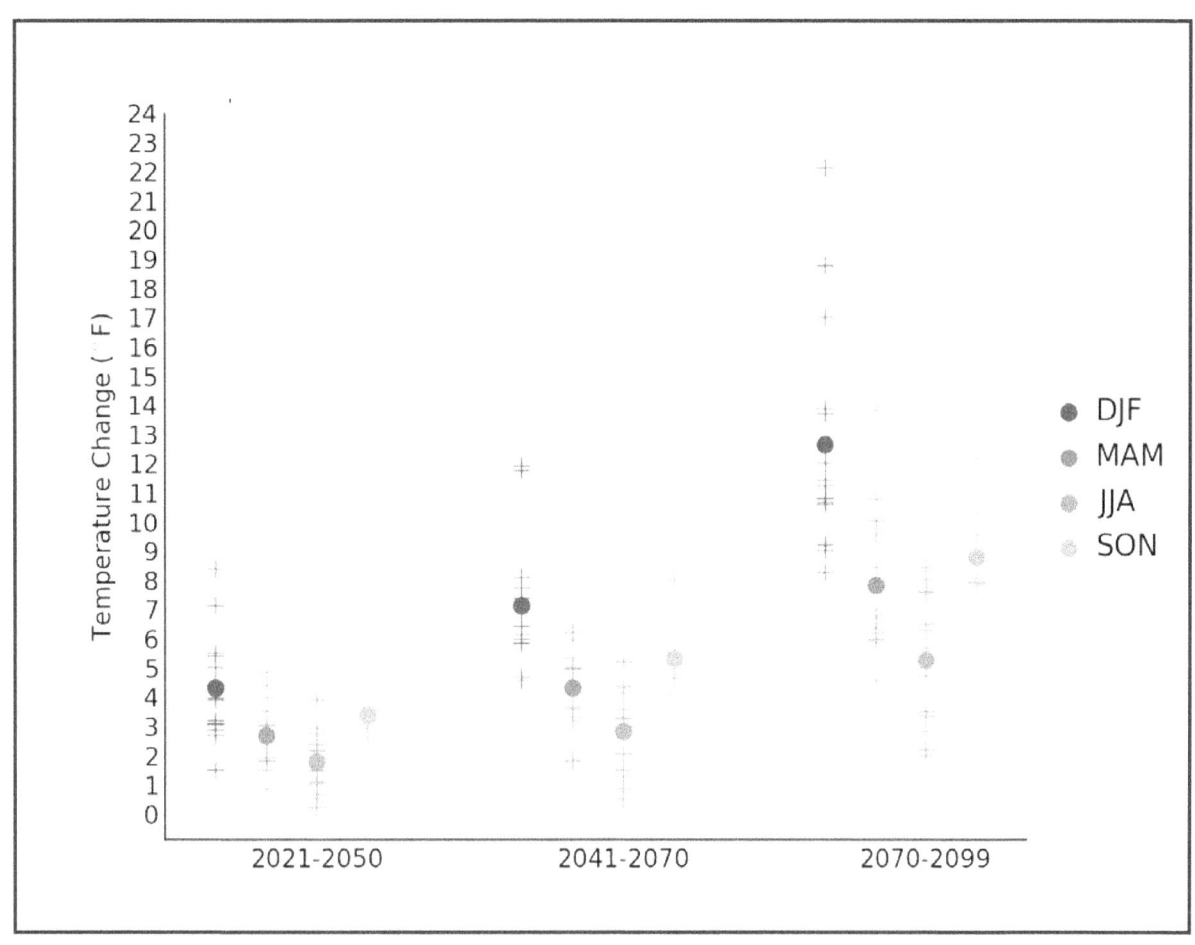

Figure 10. Simulated seasonal mean temperature change (°F) for Alaska, for each future time period (2021-2050, 2041-2070, and 2070-2099) with respect to the reference period of 1971-1999. Values are given for all 15 CMIP3 models for the high (A2) emissions scenario. The small plus signs indicate each individual model and the circles depict the multi-model means. Seasons are indicated as follows: winter (DJF, December-January-February), spring (MAM, March-April-May), summer (JJA, June-July-August), and fall (SON, September-October-November). Warming is simulated to increase with time, with changes being smallest in spring.

Table 5. Distribution of the simulated change in annual mean temperature (°F) from the 14 (B1) or 15 (A2) CMIP3 models for Alaska. The lowest, 25th percentile, median, 75th percentile and highest values are given for the high (A2) and low (B1) emissions scenarios, and for each future time period (2021-2050, 2041-2070, and 2070-2099) with respect to the reference period of 1971-1999.

Scenario	Period	Lowest	25th Percentile	Median	75th Percentile	Highest
A2	2021-2050	1.8	2.6	2.8	3.2	5.4
	2041-2070	3.5	4.2	4.7	5.1	7.7
	2070-2099	6.3	7.2	8.3	8.8	14.3
B1	2021-2050	1.6	2.2	2.9	3.4	4.4
	2041-2070	2.8	3.3	4.1	4.5	5.5
	2070-2099	3.2	3.9	4.8	5.3	8.0

This table also illustrates the overall uncertainty arising from the combination of model differences and emission pathway. For 2035, the simulated changes range from 1.7°F to 4.3°F and are almost entirely due to differences in the individual models. By 2085, the range of simulated changes has increased to 3.1°F to 11.6°F, with roughly equal contributions to the range from model differences and emission pathway uncertainties.

The preceding results have all been at the coarse resolution of the CMIP3 models. These same models have been used to produce downscaled output for Alaska at 2 km resolution, as previously described.

Figure 11 shows the downscaled annual mean temperature for 2000-2009 from Climatic Research Unit (CRU) observational data (Brohan et al. 2006), along with CMIP3 simulations for the high (A2) and low (B1) emissions scenarios for 2060-2069. The prominent role played by topography is apparent in southeastern Alaska as well as in the areas of the Alaska Range (south-central) and the Brooks Range, which separates the coastal plain to the north from the Alaskan Interior to the south.

Consistent with Fig. 8, the simulated warming generally increases from south to north and is greater in the high (A2) than in the low (B1) emissions scenario. Even in the warmer climate, however, spatial differences over scales of tens of kilometers are much larger than the changes resulting from external forcing, pointing to the inadequacy of statewide average temperatures for assessments of local and even regional impacts.

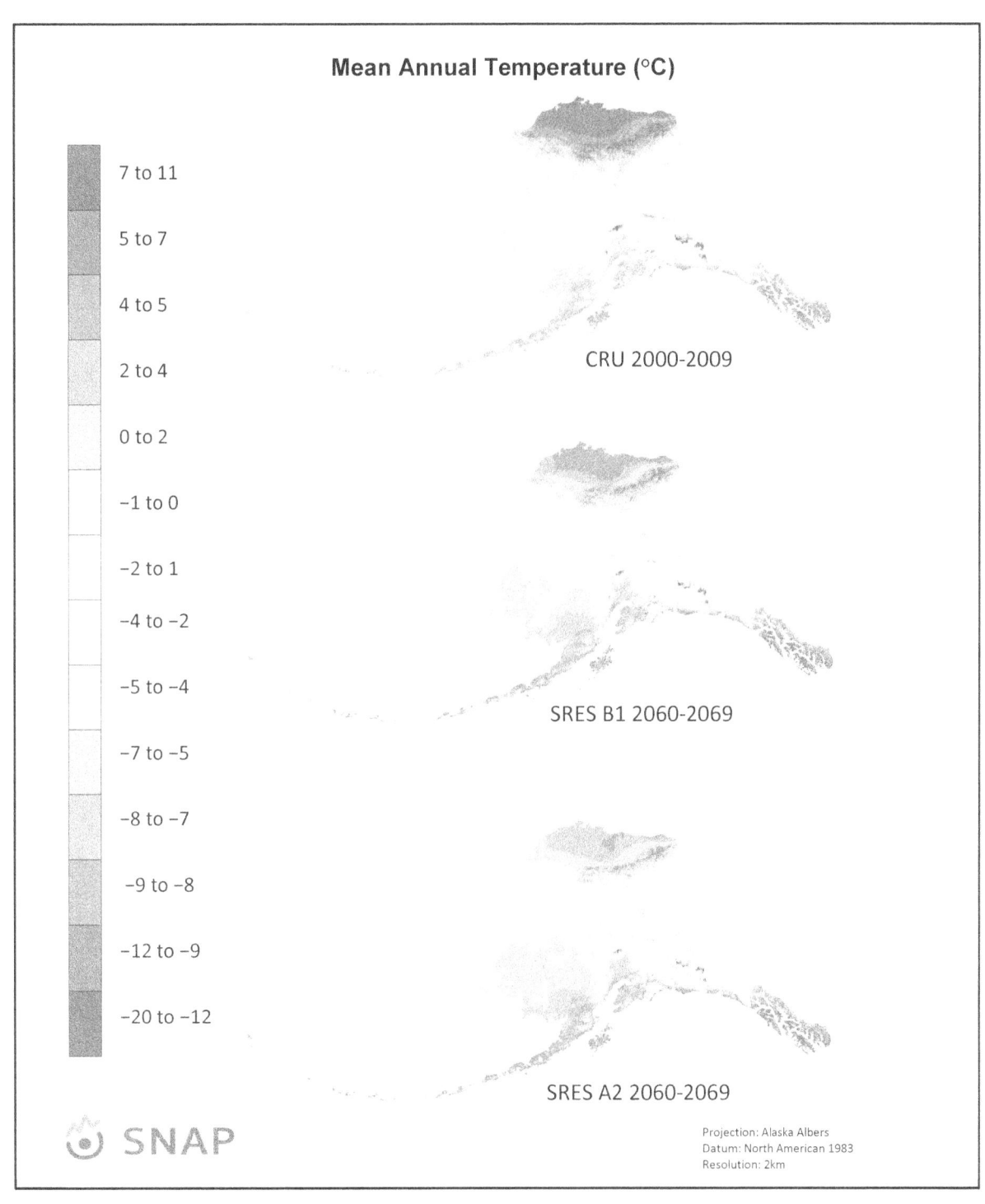

Figure 11. Decadal mean annual temperature (°C) for Alaska derived from CRU observational data for 2000-2009 (top). Simulated multi-model mean values for 2060-2069 from 5 downscaled CMIP3 global climate simulations, for the B1 (middle) and A2 (bottom) scenarios.

The impact of the scenario on the seasonal temperature changes is depicted in Figs. 12 and 13, which show the simulated CMIP3 statistically-downscaled temperature fields for winter (Dec-Feb) and summer (Jun-Aug) for the 2090-2099 timeslice. The area with temperatures close to freezing during winter is noticeably larger in the high (A2) emissions scenario, especially in south-central and southwestern Alaska (Fig. 12). The differences in the two scenarios are especially apparent in winter, when the temperatures on the North Slope are generally about -20°C in the low (B1) emissions scenario and -10° to -15°C in the A2 scenario. In summer (Fig. 13), the area with temperatures exceeding 14°C is larger in the A2 scenario. In both seasons, higher elevation areas are noticeably colder than low elevation areas, highlighting the need for downscaling when global model output is used in climate impact assessments for Alaska.

As specific examples of the statistically-downscaled CMIP3 model output for Alaska, Figs. 14-16 show the seasonal cycles of temperature for one historical reference period (1961-1990) and for different timeslices of the high (A2) and low (B1) emissions scenario simulations: 2010-2019, 2040-2049, 2060-2069 and 2090-2099. The seasonal cycles are shown for three climatically different locations: (1) Anchorage (Fig. 14) – a major population center with a generally maritime climate in the southern part of Alaska, (2) Fort Yukon (Fig. 15) – a small village with a strongly continental climate in Interior Alaska, and (3) Barrow (Fig. 16) – a village on Alaska's northern coast, where the Arctic Ocean's sea ice impinges upon the coast for about nine months of the year (although the length of the open water season is increasing). Both Fort Yukon and Barrow have largely indigenous populations that rely to varying degrees on subsistence activities.

Notable features of the temperature change at Anchorage are the pervasiveness of the warming (all calendar months) in both scenarios except for occasional single-decade decreases arising from natural variability; relatively small increases in spring (April-June) relative to those of the other seasons; the acceleration of the autumn-winter warming in both scenarios; and the greater warming in the high (A2) emissions scenario relative to B1 throughout the year. In comparison with Anchorage, Fort Yukon's simulated changes show a similar seasonality and a similar dependence on the forcing scenario. However, Fort Yukon's simulated warming is larger (consistent with Fig. 8), especially in the winter. Barrow shows an even greater warming, in excess of 20°F during October-February by the end of the century in the high (A2) emissions scenario. The seasonality of the warming is similar in the low (B1) emissions scenario, although the warming in November-February period is only about 15°F by 2090-2099 in the B1 scenario. This extremely large autumn-winter warming is partly a consequence of the loss of sea ice in the climate models, as the longer duration of open water during spring and summer allows greater oceanic storage of heat that is subsequently released to the atmosphere in autumn and early winter.

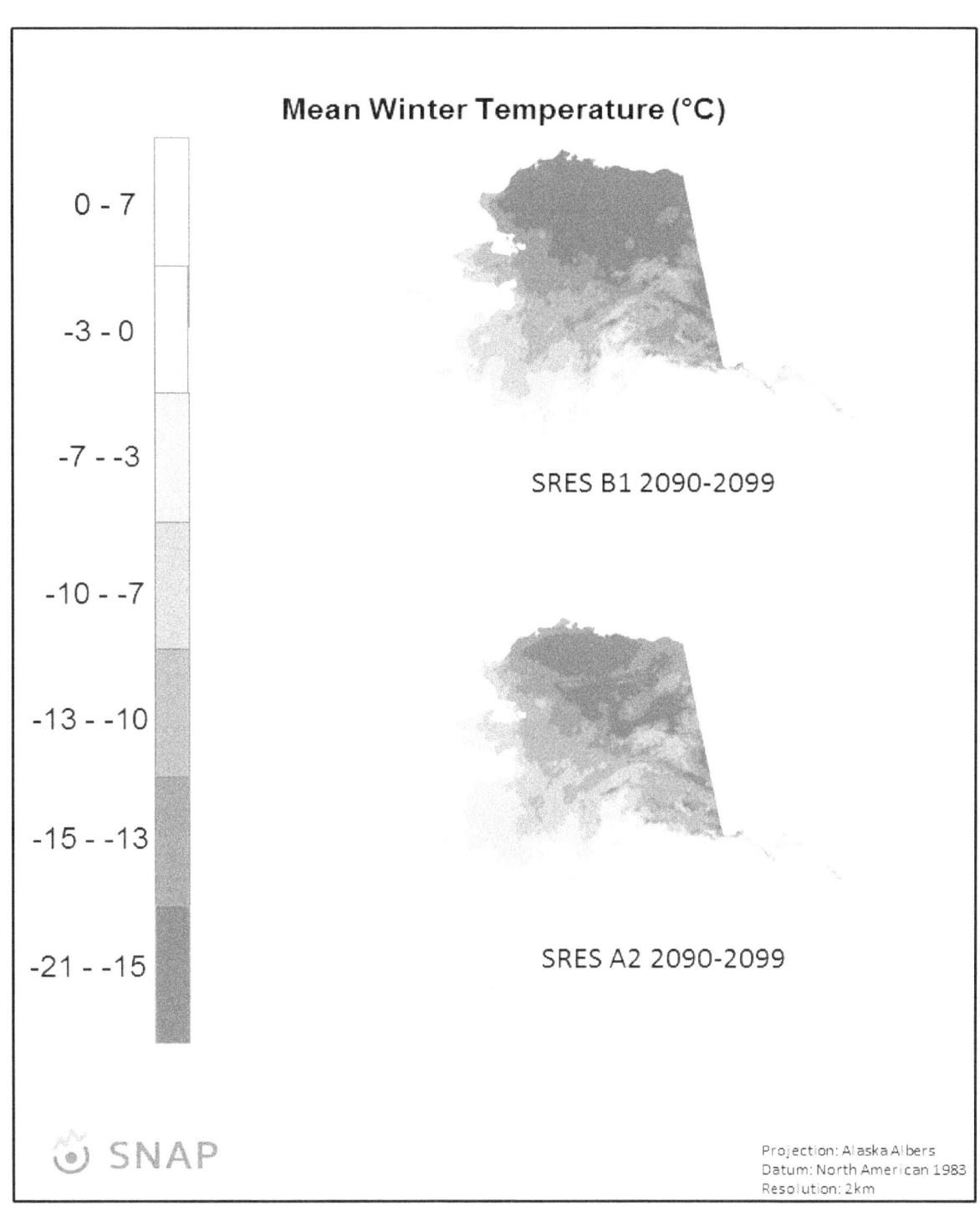

Figure 12. Simulated mean winter (Dec-Feb) temperature (°C) for Alaska for 2090-2099 for the B1 (top) and A2 (bottom) scenarios. These are downscaled multi-model means from 5 CMIP3 global climate simulations.

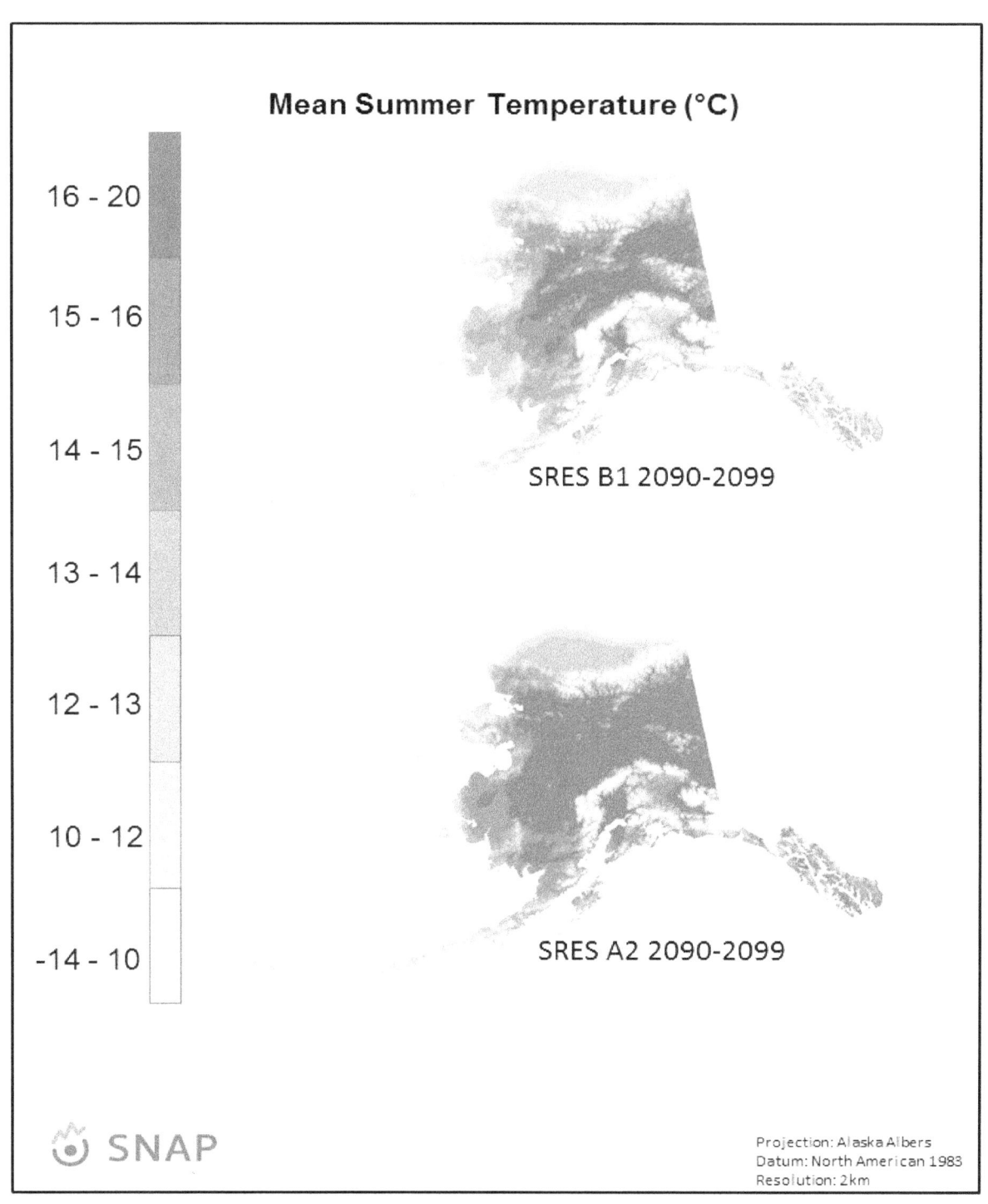

Figure 13. Simulated mean summer (Jun-Aug) temperature (°C) for Alaska for 2090-2099 for the B1 (top) and A2 (bottom) scenarios. These are downscaled multi-model means from 5 CMIP3 global climate simulations.

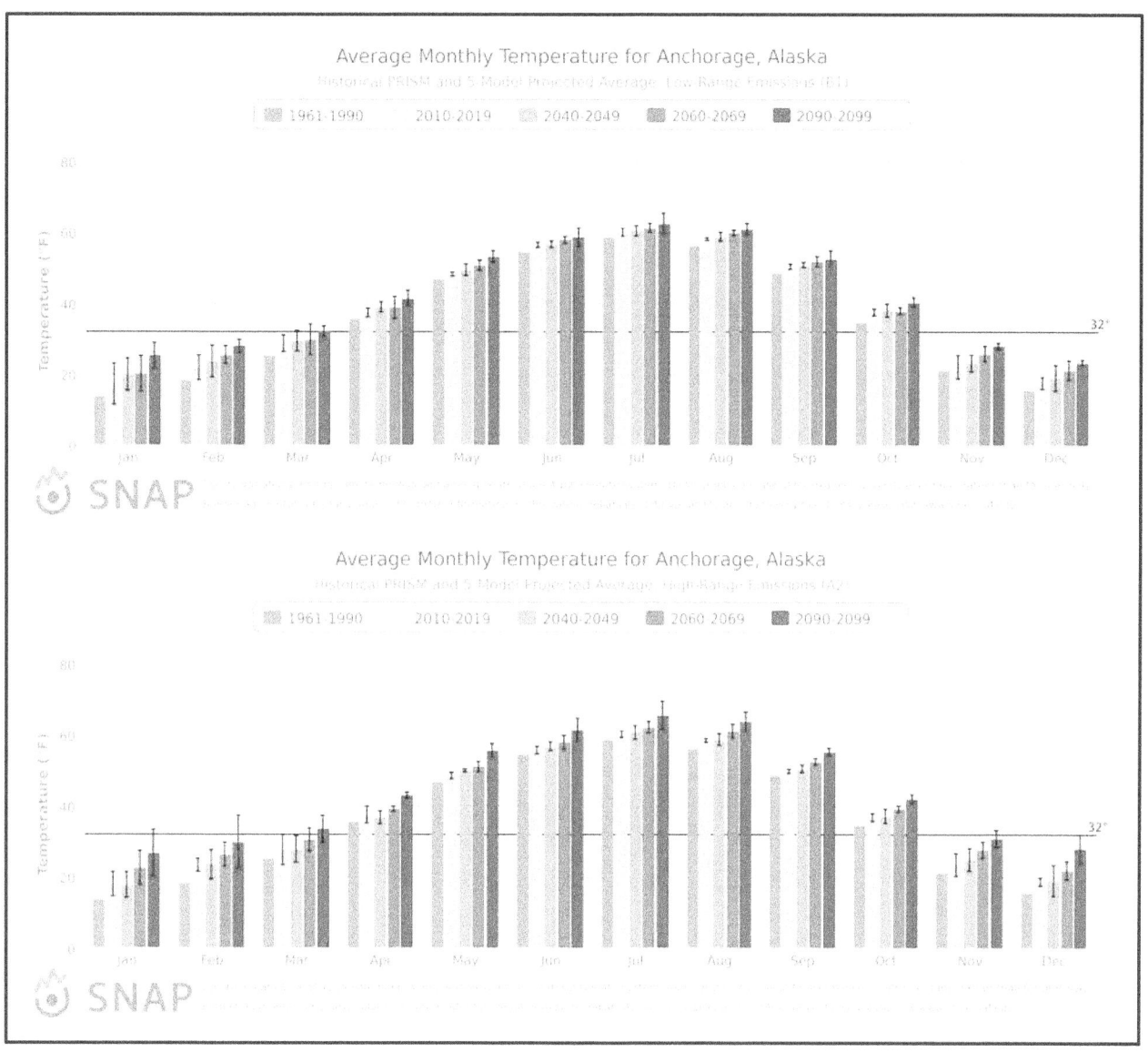

Figure 14. Simulations of decadal mean temperature (°F) by calendar month for Anchorage for the B1 (top panel) and A2 (bottom panel) scenarios. Color coding for decades is given above bar graphs. Across-model spread is indicated by thin black lines at tops of bars. These are downscaled multi-model means from 5 CMIP3 global climate simulations.

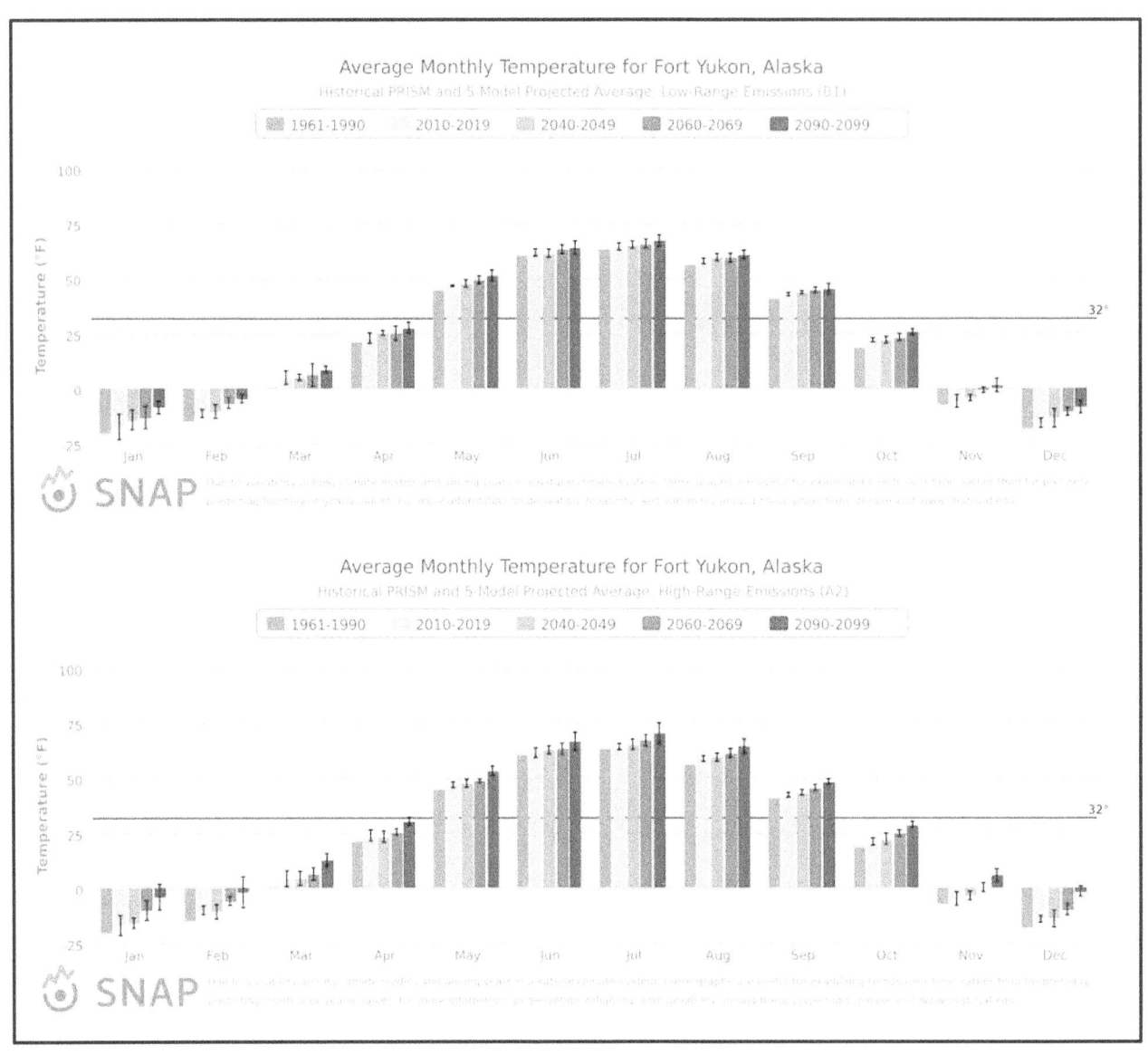

Figure 15. Simulations of decadal mean temperature (°F) by calendar month for Fort Yukon for the B1 (top) and A2 (bottom) scenarios. Color coding for decades is given above bar graphs. Across-model spread is indicated by thin black lines at tops of bars. These are downscaled multi-model means from 5 CMIP3 global climate simulations.

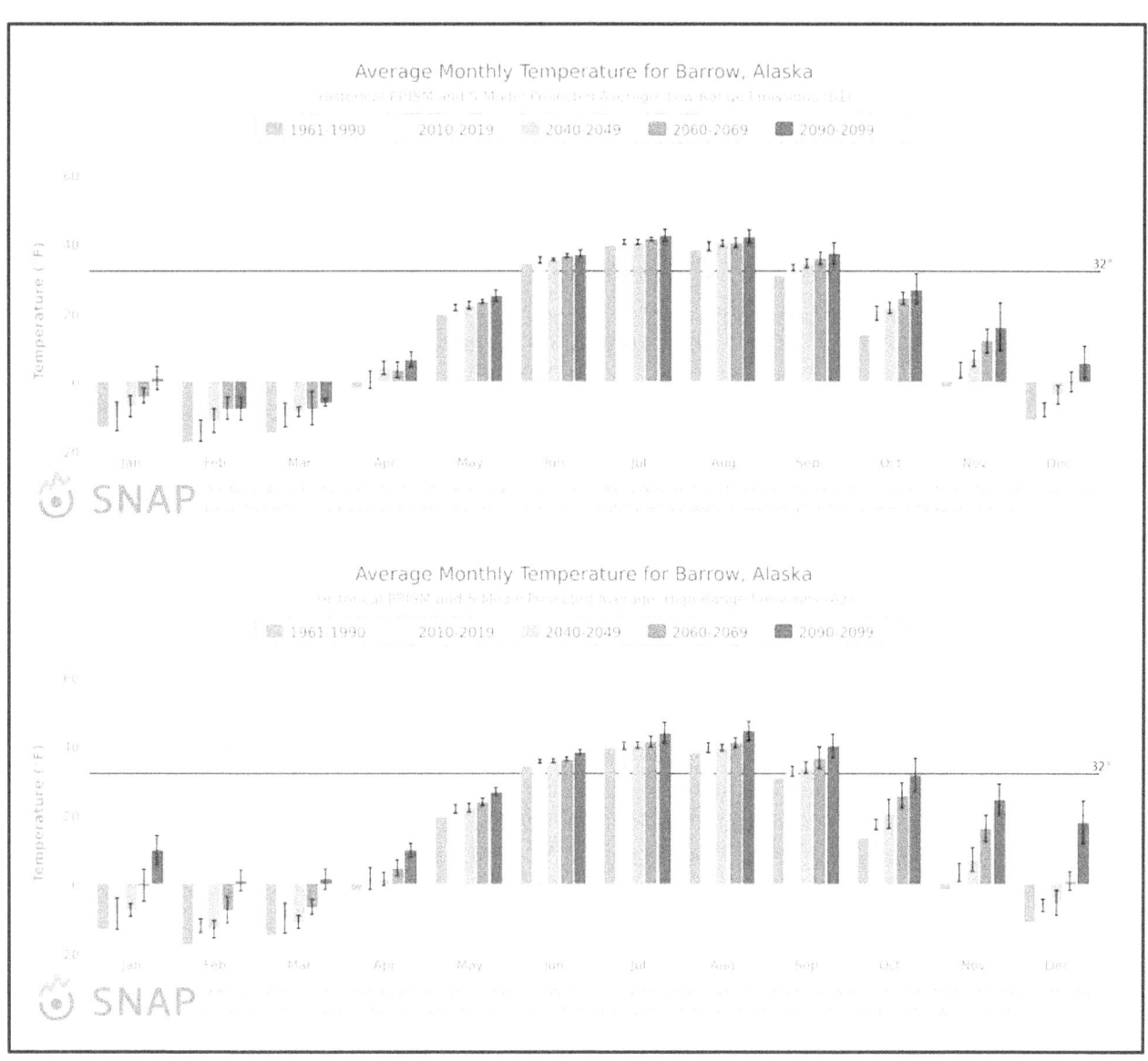

Figure 16. Simulations of decadal mean temperature (°F) by calendar month for Barrow for the B1 (top) and A2 (bottom) scenarios. Color coding for decades is given above bar graphs. Across-model spread is indicated by thin black lines at tops of bars. These are downscaled multi-model means from 5 CMIP3 global climate simulations.

Finally, Figs. 17-19 show the CMIP3 simulated change in growing season length for 2060-2069, together with the corresponding distribution for 2010-2019, all downscaled to the same 2 km resolution of the temperature fields in Figs. 11-13. Growing season length is defined here as the number of days between the spring and autumn crossings of 0°C by the daily temperature time series obtained for each year by interpolation from the monthly temperatures of that year. Increases of several weeks (15-25 days) are apparent in the southwestern and south central parts of the state. In a large portion of southwestern Alaska, the growing season lengthens to more than 200 days, a value found only along the southern coastline and subarctic islands in the present climate.

Figures 18 and 19 show IPCC model simulations of changes in thaw and freeze dates for Alaska and the surrounding area. The thaw and freeze dates were from the crossings of the 0°C threshold by the daily temperature curves, as defined in the preceding paragraph. By the last decade of the present century, the spring thaw date over much of Interior Alaska is simulated to advance by 2-3 weeks, while the autumn freeze-up is delayed by about 2 weeks. The total change on the above-freezing period is typically about 30 days over much of the state. However, a pronounced asymmetry is apparent over the waters offshore of western and northern Alaska, where the autumn freeze-up is delayed by 40-60 days. As noted earlier in the discussion of temperature changes at Barrow, the substantial delay of the first occurrence of 0°C is largely a consequence of the loss of sea ice and the seasonal (summer-autumn) storage of heat in the upper ocean.

3.4. Mean Precipitation

Figure 20 shows the spatial distribution of multi-model mean simulated differences in average annual precipitation for the three future time periods (2035, 2055, 2085) with respect to 1971-1999, for both emissions scenarios, for the 14 (B1) or 15 (A2) CMIP3 models. An increase in precipitation is simulated to occur in all cases, with the largest changes seen in the far northwest of Alaska and the smallest changes seen in the Inside Passage region. Spatial variations are again relatively small, with greater increases for the high (A2) emissions scenario than B1. The gradient of changes also increase over time, for example, precipitation differences of between 0 and 15% are indicated for the high (A2) emissions scenario in 2035, whereas for 2085 changes range from 10 to 35%. The agreement between models was once again assessed using the three categories described in Fig. 8. The CMIP3 models indicate that changes in precipitation across Alaska, for all three future time periods and both emissions scenarios, are statistically significant. The models also agree on the sign of change, with all grid points satisfying category 3, i.e. the models are in agreement that precipitation will increase throughout the state for each future time period and scenario.

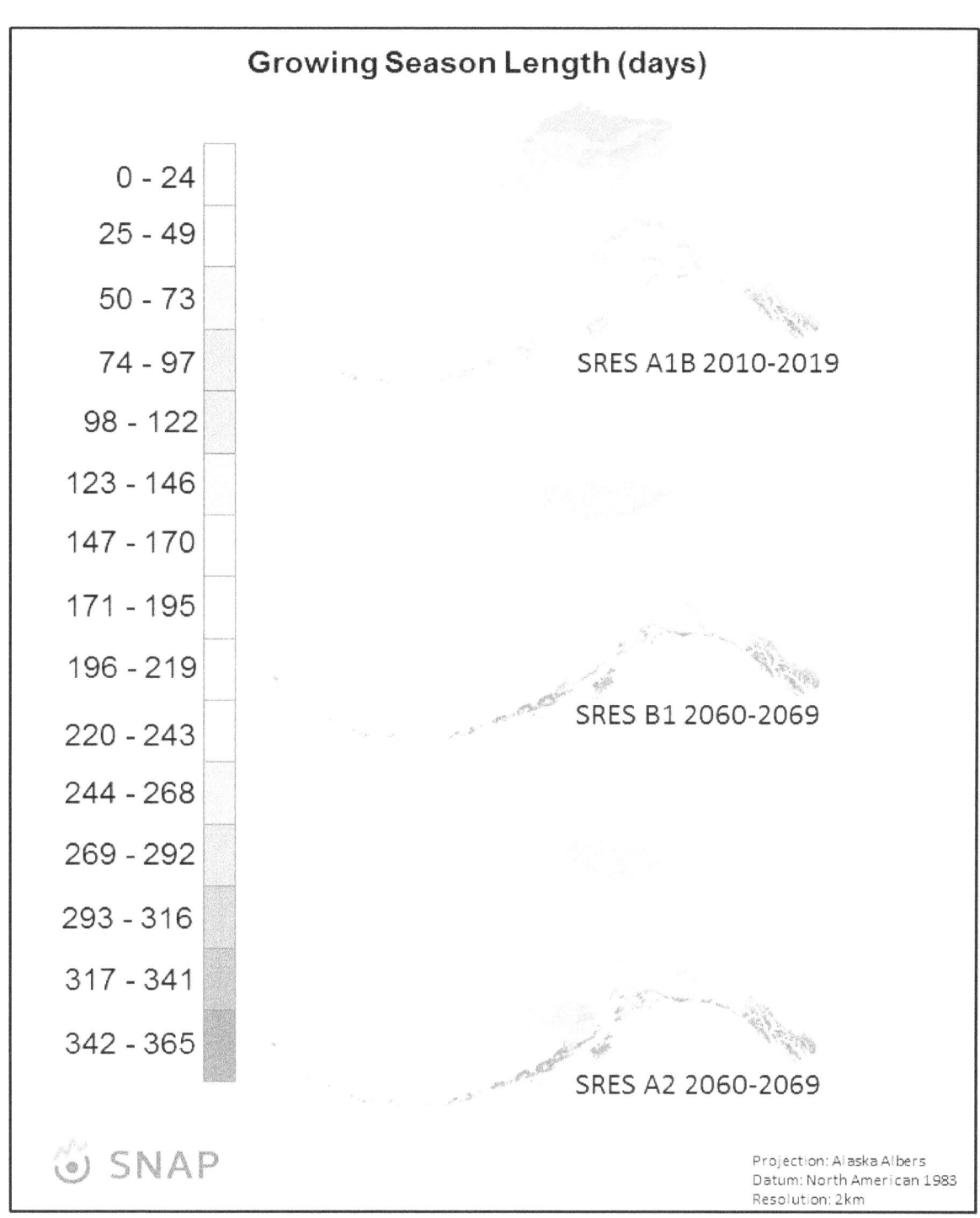

Figure 17. Decadal mean growing season length for Alaska derived from CRU observational data for 2010-2019 (top). Simulated multi-model mean values for 2060-2069 from 5 downscaled CMIP3 global climate simulations, for the B1 (middle) and A2 (bottom) scenarios.

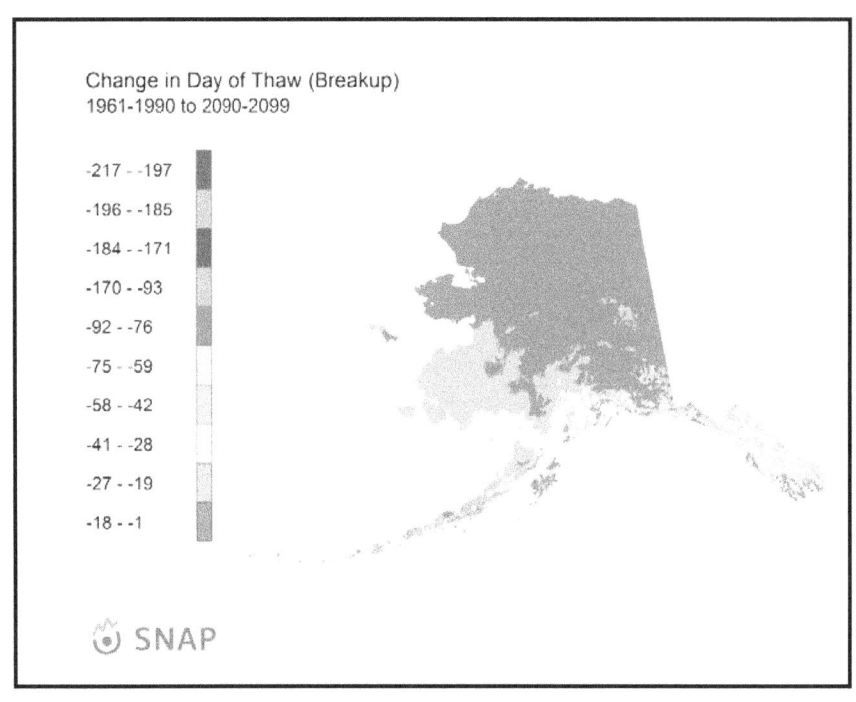

Figure 18. IPCC model simulations of change (days) in date of the final spring freeze for Alaska, for 2090-2099 relative to the reference period of 1961-1990. Negative values denote earlier thaw dates (final occurrence of 0°C). Models used were CCMA-CGCM31, MPI ECHAM5, GFDL CM2, HADCM3, and MIROC3 MEDRES.

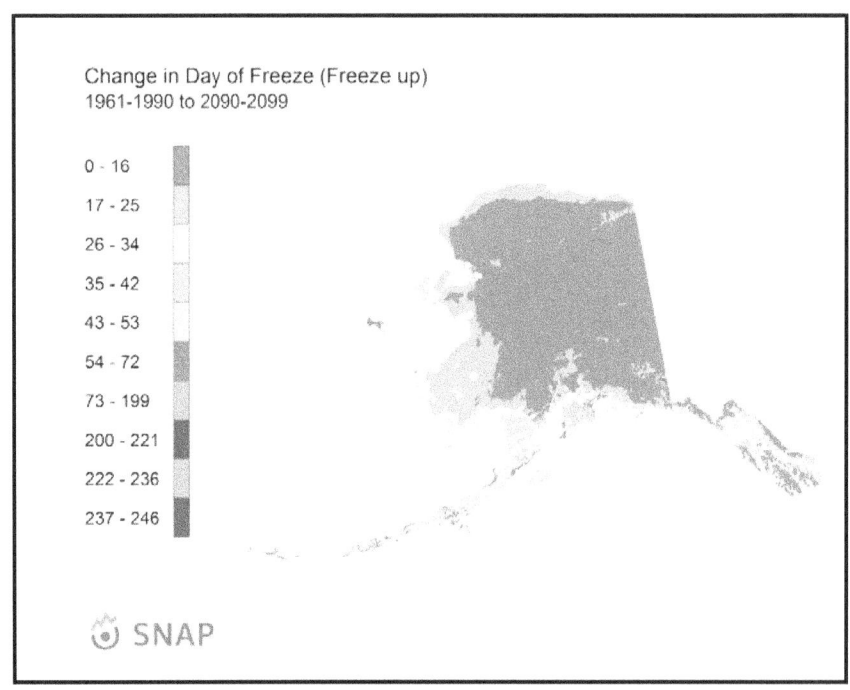

Figure 19. IPCC model simulations of change (days) in date of the first autumn freeze for Alaska, for 2090-2099 relative to the reference period of 1961-1990. Positive values denote later freeze dates (first occurrence of 0°C). Models used were CCMA-CGCM31, MPI ECHAM5, GFDL CM2, HADCM3, and MIROC3 MEDRES.

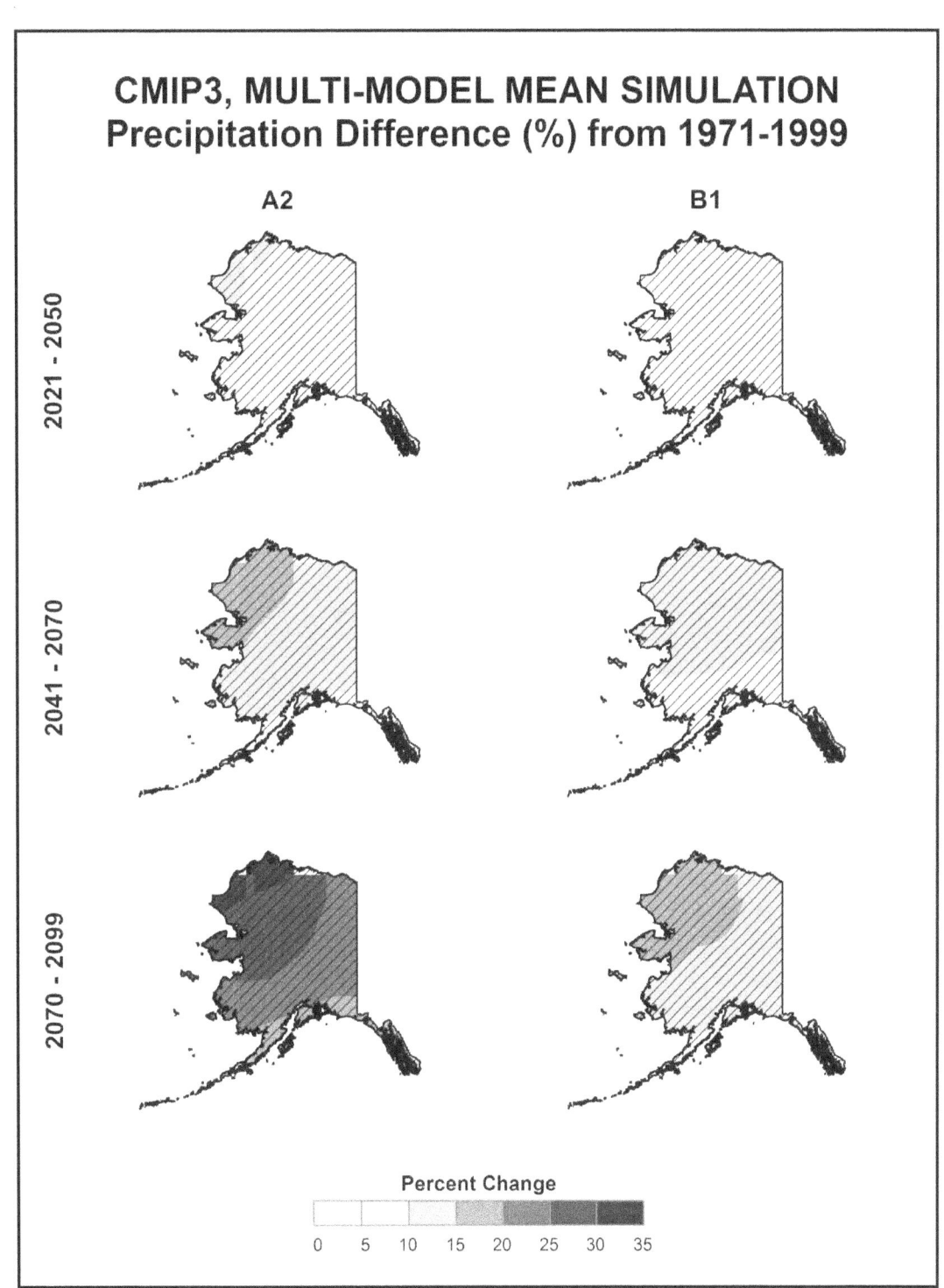

Figure 20 Simulated difference in annual mean precipitation (%) for Alaska, for each future time period (2021-2050, 2041-2070, and 2070-2099) with respect to the reference period of 1971-1999. These are multi-model means for the high (A2) and low (B1) emissions scenarios from the 14 (B1) or 15 (A2) CMIP3 global climate simulations. Color with hatching indicates that more than 50% of the models show a statistically significant change in precipitation, and more than 67% agree on the sign of the change (see text). The greatest increases in precipitation amount are simulated to occur in the far northwest of Alaska.

Table 6 shows the distribution of changes in annual mean precipitation for each future time period with respect to 1971-1999, for both emissions scenarios, across the 14 (B1) or 15 (A2) CMIP3 models. For all periods and both scenarios, the CMIP3 model simulations simulate only increases in precipitation for all time periods under both scenarios. The median values for the A2 scenario are simulated to be 8% for 2035, increasing to 25% by 2085. The inter-model range of changes (i.e. the difference between the highest and lowest model values) are between 10 and 24%. The interquartile range (the difference between the 75th and 25th percentiles) of precipitation changes across all the GCMs is less than 12%.

Table 6. Distribution of the simulated change in annual mean precipitation (%) from the 14 (B1) or 15 (A2) CMIP3 models for Alaska. The lowest, 25th percentile, median, 75th percentile and highest values are given for the high (A2) and low (B1) emissions scenarios, and for each future time period (2021-2050, 2041-2070, and 2070-2099) with respect to the reference period of 1971-1999.

Scenario	Period	Lowest	25th Percentile	Median	75th Percentile	Highest
A2	2021-2050	1	6	8	11	15
	2041-2070	5	10	12	15	22
	2070-2099	11	17	25	28	35
B1	2021-2050	1	5	7	11	12
	2041-2070	5	7	10	13	15
	2070-2099	7	11	14	16	10

Table 7 shows the distribution of changes in seasonal mean precipitation across the 14 (B1) or 15 (A2) CMIP3 models, between 2070-2099 and 1971-1999 for both emissions scenarios. The only season for which any model simulates a decrease in precipitation is spring (under the B1 scenario). The range of model-simulated changes is quite large. For example, in the high (A2) emissions scenario, the simulated change in winter precipitation varies from +10% to +48%. In the low (B1) emissions scenario, the range of precipitation changes is generally smaller, with the largest simulated increase being 23% in winter. The central feature of the results in Table 7 is the large uncertainty in seasonal precipitation changes.

It is important to note that an increase in overall precipitation amount does not necessarily imply an increase in water availability. As temperatures rise, growing season for various types of vegetation will increase, thereby increasing the overall water uptake by plants. Similarly, warming temperatures will increase evaporation as well as the atmosphere's holding capacity for water vapor. Thus, while model predictions may indicate a general increase in precipitation amount, this may not result in increased water availability (see Figure 10.12 in IPCC 2007a).

Table 7. Distribution of the simulated change in seasonal mean precipitation (%) from the 14 (B1) or 15 (A2) CMIP3 models for Alaska. The lowest, 25th percentile, median, 75th percentile and highest values are given for the high (A2) and low (B1) emissions scenarios, and for the 2070-2099 time period with respect to the reference period of 1971-1999.

Scenario	Period	Season	Lowest	25th Percentile	Median	75th Percentile	Highest
A2	2070-2099	DJF	10	21	30	38	48
		MAM	8	18	22	26	40
		JJA	5	14	21	25	30
		SON	11	17	24	30	35
B1	2070-2099	DJF	6	10	19	20	23
		MAM	-4	9	12	14	21
		JJA	7	10	12	14	19
		SON	6	10	11	17	23

Figure 21 shows the change in annual mean precipitation for each future time period with respect to 1971-1999, for both emissions scenarios, averaged over the entire Alaska region for the 14 (B1) or 15 (A2) CMIP3 models. Both the multi-model mean and individual model values are shown. All the simulated multi-model mean changes in precipitation indicate increases. For the high (A2) emissions scenario, the CMIP3 models simulate changes in precipitation of from 8 to 23%, increasing for each future time period. For the low (B1) emissions scenario, the values are comparable to those for A2 in 2035 and 2055, but considerably lower for 2085.

The range of individual model changes in Fig. 21 is large compared to the differences in the multi-model means, as also illustrated in Table 6. In fact, for both emissions scenarios, the individual model range is much larger than the differences in the CMIP3 multi-model means between time periods.

Figure 22 shows the change in seasonal mean precipitation for each future time period with respect to 1971-1999, for the high (A2) emissions scenario, averaged over the entire Alaska region for the 15 CMIP3 models. Again, both the multi-model mean and individual model values are shown. The simulated multi-model mean precipitation changes are largest in winter, ranging from +10 to +28%. For the other three seasons, the mean simulated changes are also positive and increase over time. As was the case for the annual totals in Fig. 21, the model ranges in Fig. 22 are large compared to the multi-model mean differences. This illustrates the large uncertainty in the precipitation estimates using these simulations.

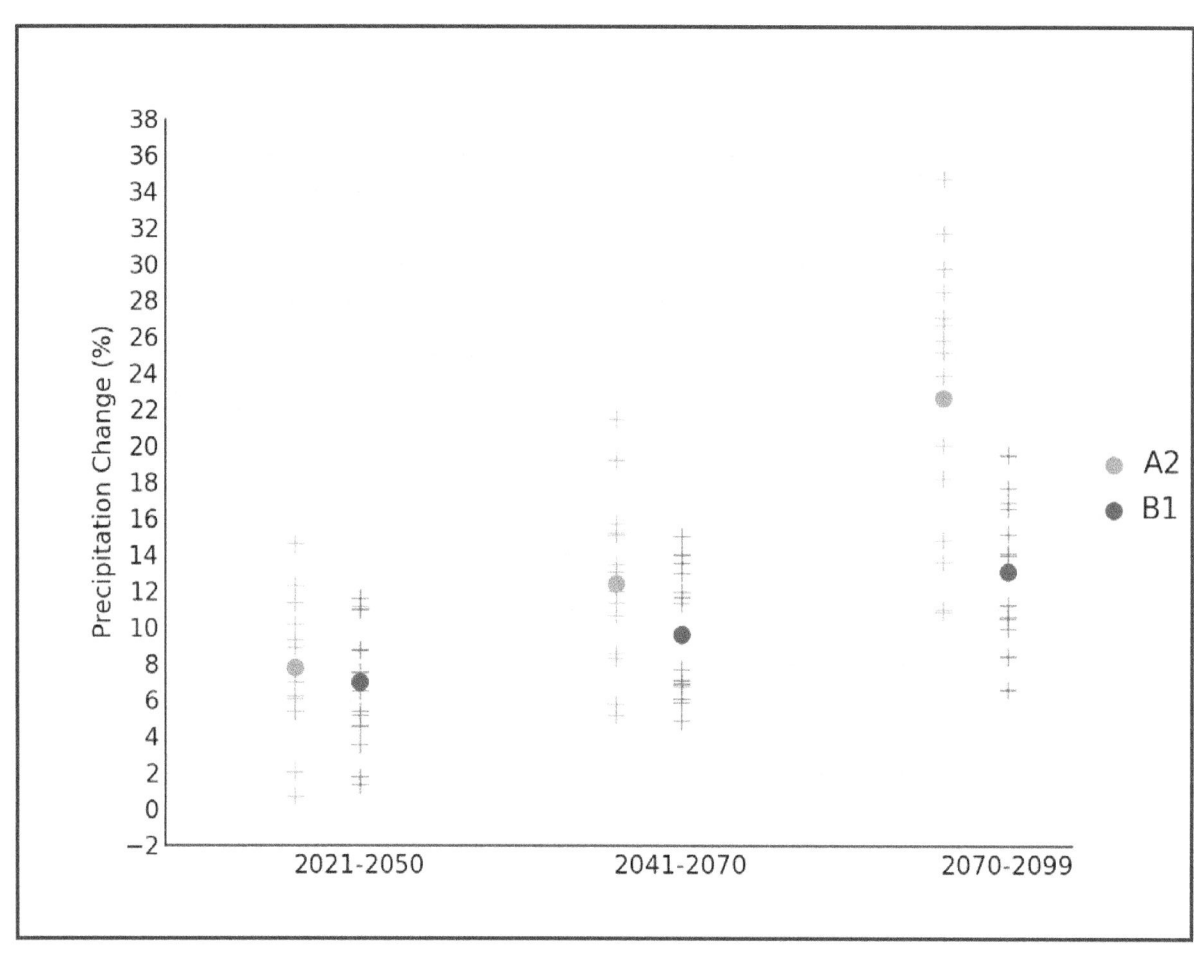

Figure 21. Simulated annual mean precipitation change (%) for Alaska, for each future time period (2021-2050, 2041-2070, and 2070-2099) with respect to the reference period of 1971-1999. Values are given for the high (A2) and low (B1) emissions scenarios for the 14 (B1) or 15 (A2) CMIP3 models. The small plus signs indicate each individual model and the circles depict the multi-model means. The ranges of model-simulated changes are very large compared to the mean changes and to differences between the A2 and B1 emissions scenarios.

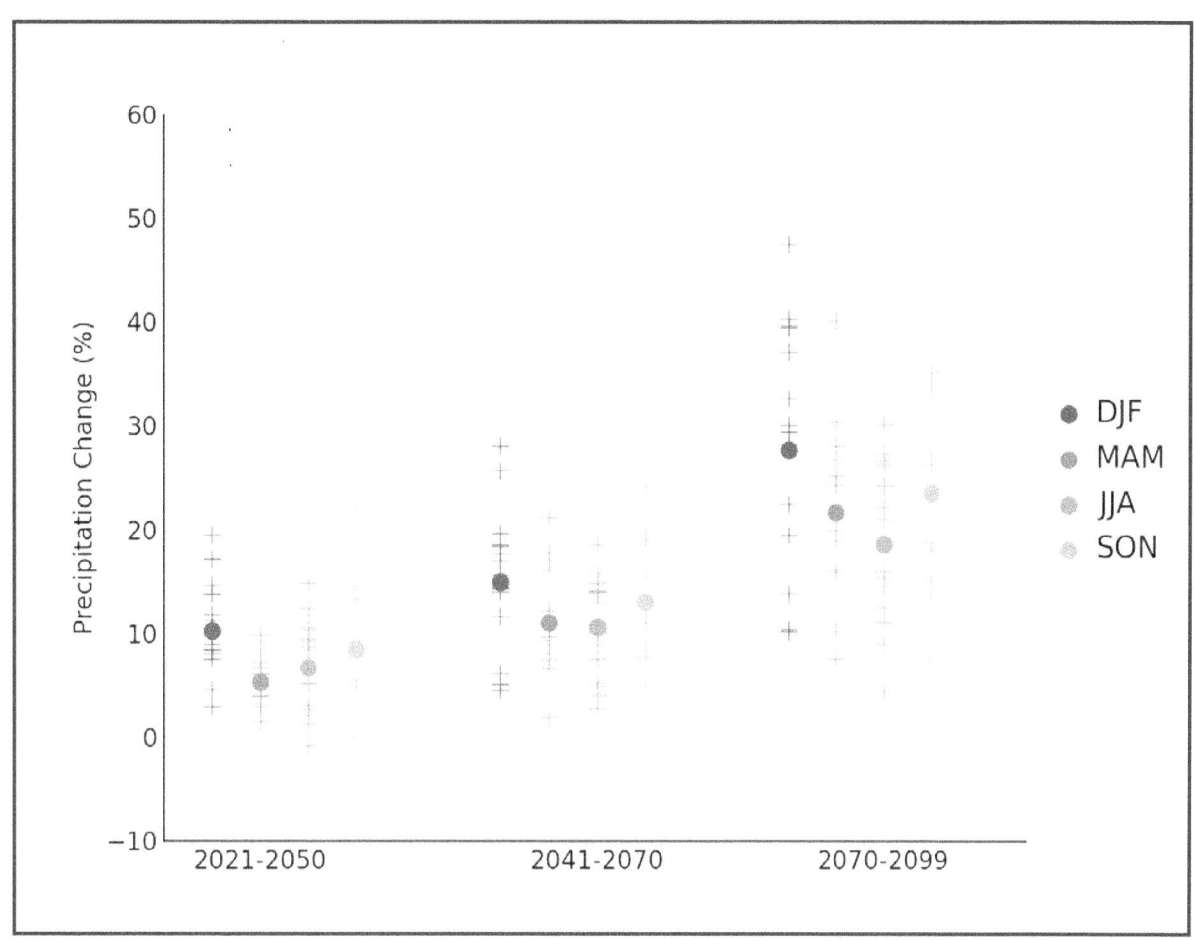

Figure 22. Simulated annual mean precipitation change (%) for Alaska, for each future time period (2021-2050, 2041-2070, and 2070-2099) with respect to the reference period of 1971-1999. Values are given for all 15 CMIP3 models for the high (A2) emissions scenario. The small plus signs indicate each individual model and the circles depict the multi-model means. Seasons are indicated as follows: winter (DJF, December-January-February), spring (MAM, March-April-May), summer (JJA, June-July-August), and fall (SON, September-October-November). The ranges of model-simulated changes are large compared to the mean changes and to differences between the seasons.

Precipitation in Alaska shows a spatially complex pattern, largely as a result of the state's major topographic features: the coastal mountains of the Southeast, the Alaska Range in the south-central region, and the Brooks Range in the north. Figure 23 shows downscaled fields of annual mean precipitation for the present climate of 2000-2006 (from CRU observational data) and for 2060-2069 (CMIP3 simulations) under the high (A2) and low (B1) emissions scenarios. As for Fig. 11, the simulations are composited over simulations by the five CMIP3 models that best capture the seasonal cycles of key climate variables in Alaska over the past several decades. All panels of Fig. 23 show the huge range of annual precipitation amounts, with less than 20 cm annually on the North Slope and more than several meters in the Southeast. Both scenarios indicate general increases by the decade of the 2060s. The absolute increases are largest in the Southeast, while the percentage increases are largest in the western and northern parts of the state.

In order to illustrate the seasonality of the changes of precipitation locally, Figures 24-26 show decadal timeslices of downscaled CMIP3 calendar-month mean precipitation for the same locations for which temperature simulations were shown earlier (Figs. 14-16): Anchorage near the southern coast, Fort Yukon in the Interior, and Barrow on the northern coast. There are several noteworthy features of the precipitation amounts and simulated changes in Figs. 24-26. First, cold-season precipitation amounts are much larger at Anchorage than at the other two stations (note the different scales used for precipitation amounts). Second, the amounts are more similar across the stations in summer, when much of the precipitation is convective (especially at Fort Yukon). However, the variation among models is greatest for the summer and autumn months (extending into winter for Barrow). Third, precipitation amounts generally increase in all calendar months at all stations in both scenarios, although each station shows a few examples of future decadal decreases due to natural variability. The fact that such decreases occur even though the bars in the figures represent five-model means points to the importance of natural variability on decadal timescales. Also, the high (A2) emissions scenario shows an acceleration of the increase of precipitation in the final decade, especially in summer and autumn. The possible role of increased open water (loss of sea ice) in these large increases merits investigation, especially since a similar type of behavior was apparent in the corresponding temperature plots. The prominence of this acceleration in A2 but not in B1 results also merits closer examination, as it may point to an impact of mitigation if mitigation actions indeed shape future greenhouse gas concentrations.

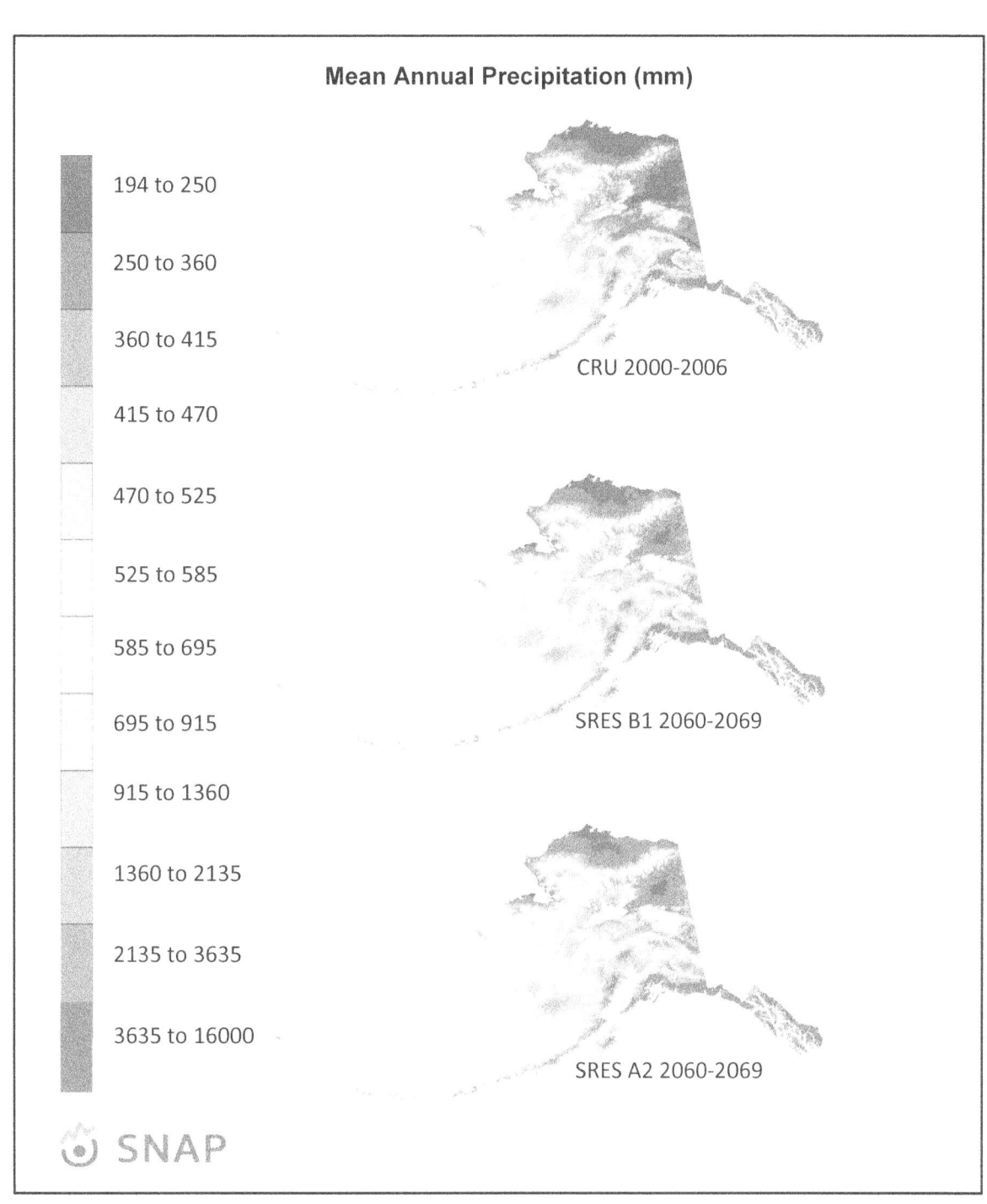

Figure 23. Decadal mean annual precipitation (mm) for Alaska derived from CRU observational data for 2000-2006 (top). Simulated multi-model mean values for 2060-2069 from 5 downscaled CMIP3 global climate simulations, for the B1 (middle) and A2 (bottom) scenarios.

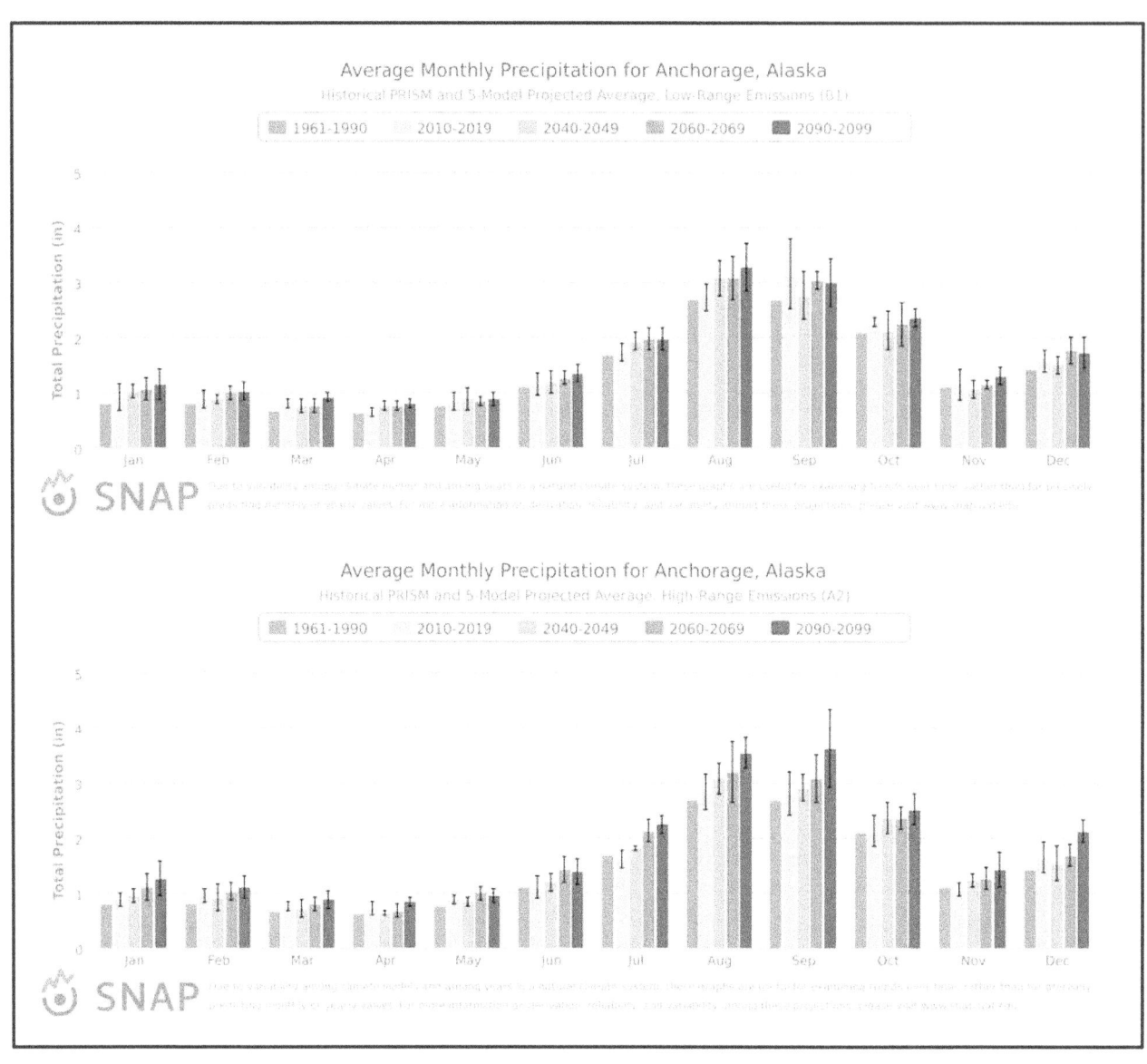

Figure 24. Simulations of decadal mean precipitation (inches) by calendar month for Anchorage for the B1 (top) and A2 (bottom) scenarios. Color coding for decades is given above bar graphs; deeper blues indicate increasing range of simulations. Across-model spread is indicated by thin black lines at tops of bars. These are downscaled multi-model means from 5 CMIP3 global climate simulations.

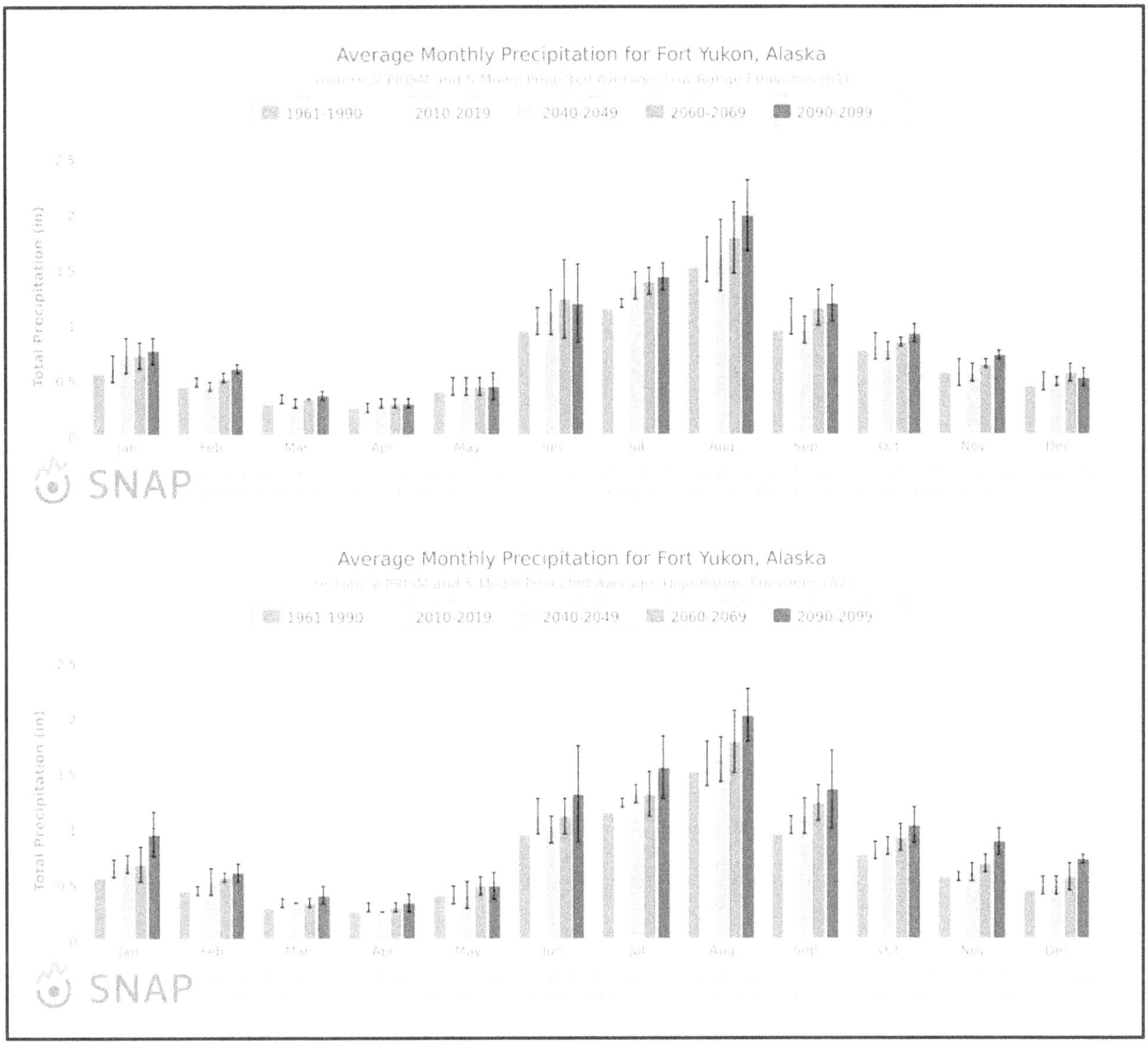

Figure 25. Simulations of decadal mean precipitation (inches) by calendar month for Fort Yukon for the B1 (top) and A2 (bottom) scenarios. Color coding for decades is given above bar graphs; deeper blues indicate increasing range of simulations. Across-model spread is indicated by thin black lines at tops of bars. These are downscaled multi-model means from 5 CMIP3 global climate simulations.

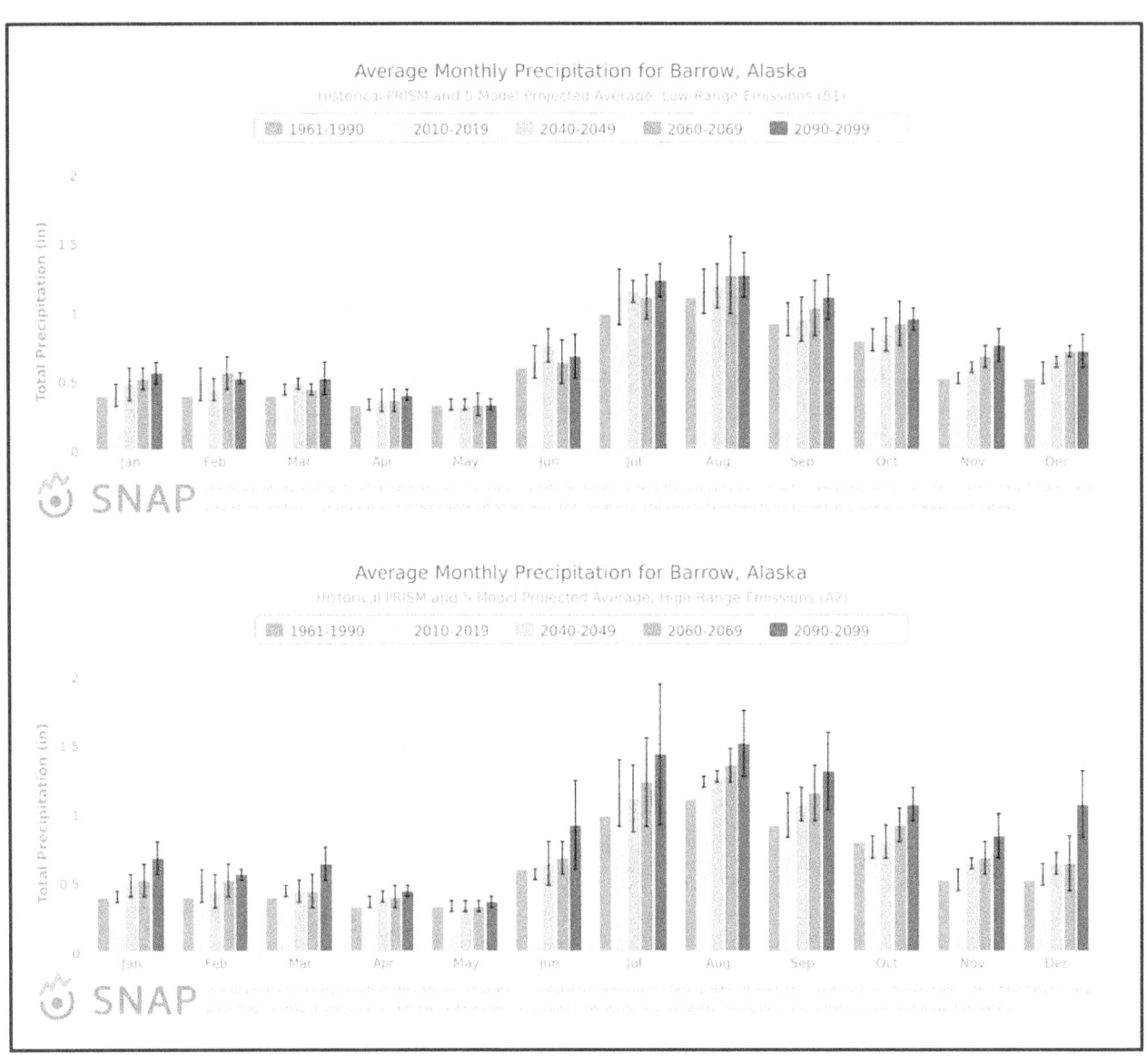

Figure 26. Simulations of decadal mean precipitation (inches) by calendar month for Barrow for the B1 (top) and A2 (bottom) scenarios. Color coding for decades is given above bar graphs; deeper blues indicate increasing range of simulations. Across-model spread is indicated by thin black lines at tops of bars. These are downscaled multi-model means from 5 CMIP3 global climate simulations.

3.5. Permafrost

Because permafrost thaw results in the settling and/or slumping of soil as well as altered hydrology, it is one of the serious impacts of a warming climate in Alaska. Climate model output has been used to drive permafrost models in order to capture the degradation of permafrost under greenhouse-driven climate simulations based on the SRES scenarios[6]. In the examples shown here, the University of Alaska's Geophysical Institute Permafrost Laboratory (GIPL) permafrost model has been forced by downscaled (2 km) temperature and precipitation (snowfall) output from the high (A2) and low (B1) emissions scenario simulations of five CMIP3 models described in Section 2.

Figure 27 shows the CMIP3 downscaled 2 km resolution annual mean ground temperatures at 1 meter depth from the high (A2) and low (B1) emissions scenario simulations for three timeslices: 2001-10, 2041-50 and 2091-2100. An annual mean ground temperature of 0°C (32°F) at 1 meter depth is an effective indicator of stable permafrost ($T < 0°C$) or degrading/non-permafrost ($T > 0°C$). In the figures below, the 1 meter temperatures are color-coded in blue for temperatures below 0°C and yellow-to-red for temperatures above 0°C. It is apparent from the figures that the area of above-freezing ground expands dramatically during the 21st century under both scenarios, especially in the second half of the century. The area of permafrost degradation (change from blue to yellow/orange) covers much of Interior Alaska in the high (A2) emissions scenario. The thaw is more discontinuous in the B1 simulation, although patches of thaw extend as far north as the Brooks Range even under the low (B1) emissions scenario. Because permafrost thaw has severe consequences for infrastructure (buildings, pipelines, airports, roads), the changes depicted in Fig. 27 represent major risks of climate change in Alaska (Larsen et al. 2008).

3.6. Sea Ice

Recent climate variability in the Arctic has resulted in significant sea ice changes in the Bering and Chukchi Seas. These changes are most pronounced in the late summer and early fall, defined as July through October. Meier et al. (2007) studied regional variations in Arctic sea ice extent and found that Bering Sea ice decreased 39 to 43 percent in July and October from 1979-2006. In the Chukchi region, July through October sea ice extent declined by 24 to 47 percent. The overall downward trend of Arctic sea ice can be explained from increasing air and ocean temperatures and changing atmospheric and ocean circulation (Stroeve et al. 2011). An anti-cyclonic regime has dominated the Arctic Ocean for over a decade, which results in greater ice movement of older, thicker ice towards the Atlantic Ocean. The remaining thinner ice is more vulnerable to melting during the summer months. In addition, the inflow of heat from the Pacific to the Arctic Ocean through the Bering Strait has increased during the past decade, resulting in a situation in which the greatest loss of summer sea ice has been in the Alaskan/East Siberian sector. It should be noted, however, that the most recent winters (2008-2011) have seen sufficient refreezing to reduce the magnitude of the negative ice extent anomalies during winter in the Bering Sea. This tendency for the reduction of sea ice to be greater in summer than in winter is consistent with greenhouse-driven climate model simulations, which simulate greater reductions of ice extent in summer than in winter.

---

[6] Refers to the IPCC Special Report on Emissions Scenarios (IPCC 2000).

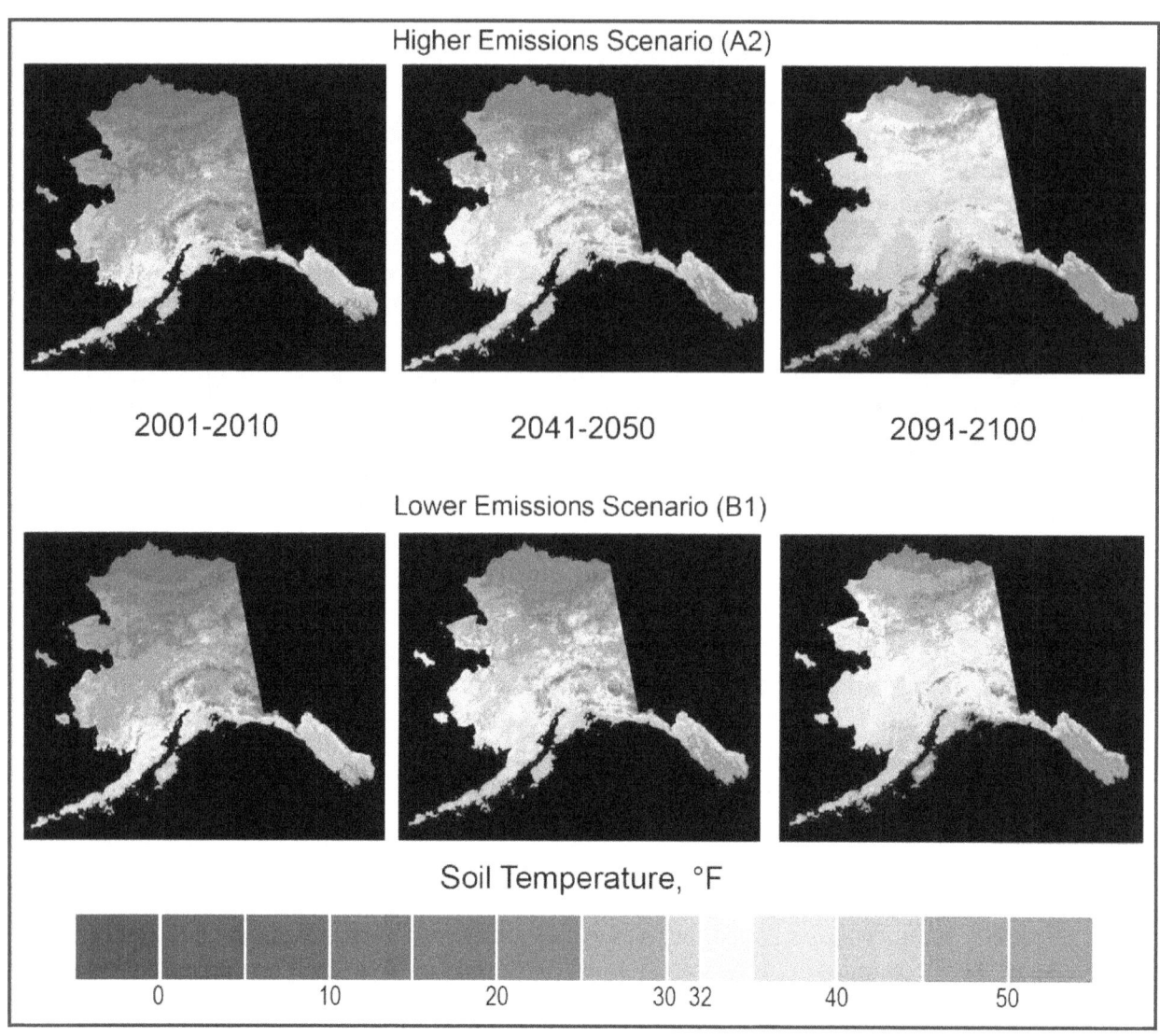

Figure 27. Annual mean ground temperatures at 1-meter depth for 2001-10, 2041-50, and 2091-2100. Fields are from GIPL permafrost model simulations driven by downscaled CMIP3 climate model output for the B1 (bottom panels) and A2 (top panels) scenarios. As indicated by color bars, blue shades represent temperatures below 32°F (0°C) and yellow-to-red shades represent temperatures above 32°F (0°C). Figure provided by S. Marchenko and the Geophysical Institute Permafrost Laboratory, University of Alaska, Fairbanks.

On a regional basis, climate models simulate large declines in sea ice extent in the Alaskan region. Figure 28 shows the simulations from a subset of CMIP3 models that were found to produce the most realistic simulations of recent sea ice conditions in the Alaskan subregions. Summer sea ice in the Chukchi Sea disappears between 2030 and 2050 in some models, while winter sea ice in the Bering and Chukchi Seas decrease by more than 50 percent by the end of the century (Overland and Wang 2007). It should be noted that the simulations summarized in Fig. 28 are based on the A1B scenario; the rates of ice loss are larger (smaller) in the A2 (B1) emissions scenario.

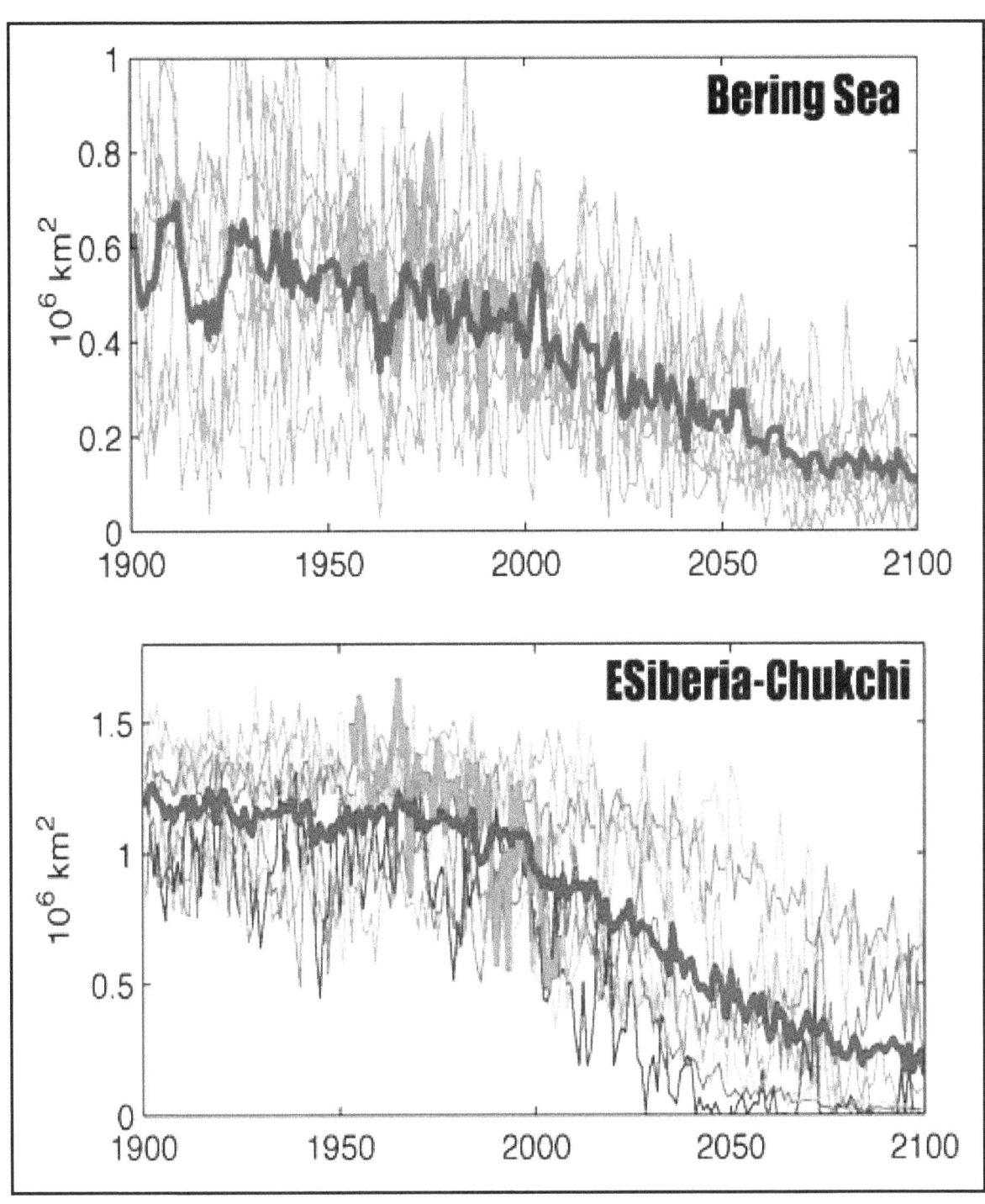

Figure 28. Winter (March-April average) sea ice area (10^6 km^2) in the Bering Sea, and summer (August-September average) sea ice area in the East Siberia-Chukchi Seas from 1900 to 2100, as simulated by different CMIP3 models (thin colored lines). The thick blue line is a multi-model mean, and the thick red line shows recent observational data. Republished with permission of the American Geophysical Union, from Overland and Wang (2007); permission conveyed through Copyright Clearance Center, Inc.

Changes in Arctic sea ice have significant implications for marine access. For example, the rates of ice reduction in the CMIP3 models imply a lengthening of the open water season in the Bering Strait by 2 to 3 months by the late 21st century. This increase is consistent with the changes in break-up and freeze-up dates shown in Figs. 18 and 19. In 2009, the Arctic Council reviewed the physical, economic, social, and political concerns related to increased Arctic marine access, reviewing sea ice simulations from the Fourth IPCC Assessment. This report, the Arctic Marine Shipping Assessment (AMSA), found that increased Arctic accessibility will likely result in greater marine access and resource development. The Northwest and Northeast Passages are expected to become regular shipping lanes for increasing periods of the year as the century progresses. With its longer ice-free season length, the Bering Strait could become a major throughput for marine vessels seeking the Northwest and Northeast Passages. This will require new infrastructure in Alaska to handle increased marine activity. In view of the likelihood that seasonal sea ice will continue to form during winter, the AMSA recommended a greater presence of icebreaking capable vessels in the Arctic in order to maximize accessibility (AMSA 2009).

4. SUMMARY

The primary purpose of this document is to provide physical climate information for potential use by the authors of the Third National Climate Assessment report. The document contains two major sections. One section summarizes historical conditions in Alaska, focusing primarily on trends in key climate conditions and identifying climate factors that are important in the region. The core data sets used to examine temperature and precipitation were those of the National Climatic Data Center's First-Order Surface Weather Observing Stations.

The second section summarizes climate model simulations for two scenarios of the future path of greenhouse gas emissions: the IPCC SRES high (A2) and low (B1) emissions scenarios. These simulations are based on analyses from Phase 3 of the Coupled Model Intercomparison Project (CMIP3) simulations. Analyses of the simulated future climate are provided primarily for periods of 2021-2050, 2041-2070, and 2070-2099, with changes calculated with respect to an historical climate reference period of 1971-1999. The resulting climate conditions are to be viewed as scenarios, not forecasts, and there are no explicit or implicit assumptions about the probability of occurrence of either scenario. The basis for these climate scenarios (emissions scenarios and sources of climate information) were considered and approved by the National Climate Assessment Development and Advisory Committee.

Some key characteristics of the historical climate include:

- Climatic phenomena that have major impacts on Alaska include increasing temperature, thawing permafrost, coastal storms, ocean acidification, floods and drought.

- Temperature anomalies (annual departures from the long-term average) were mostly negative (cool) in the mid-20th century, shifting to mostly positive (warm) around 1976 when the Pacific Decadal Oscillation (PDO) underwent a phase shift from negative to positive. It is apparent from regional and seasonal analyses that most of the warming has occurred in winter and spring for all regions of Alaska and that the Interior region has experienced the greatest overall warming.

- Extreme temperatures display similar regional and seasonal variation to those of mean temperatures in Alaska. Increases in the number of extreme warm events and decreases in the number of extreme cold events are greatest in spring. Over the past few decades, the observed decrease in frequency of extreme cold events has become more pronounced. Temporal patterns again coincide with the phase shift of the PDO around 1976.

- There is some indication that frost-free and growing seasons are lengthening. While comprehensive analyses of historical trends are lacking, evaluations have been performed for individual stations having long records.

- Long-term trends in average annual precipitation show nearly average amounts from 1949 to 1965, followed by about 15 years of below average totals. Recent decades have seen precipitation amounts largely above average in Alaska.

- No significant change in the overall frequency of strong storm events is detectable, though the northern and northwestern coasts have seen a significant increase in the occurrence of strong storms when a protective sea ice cover is not present during summer and autumn. Overall storm frequencies show little trend south of the Bering Strait, where loss of sea ice has been less.

- The occurrence of fires in Interior Alaska has increased under the influence of warmer spring temperatures, earlier snow melt, longer growing seasons, and deeper permafrost active layers. The past several years have also seen unprecedented fire occurrence on the tundra of northern and western Alaska associated with sea ice loss.

- Sea ice extent at the end of each of the past six summers (2007-2012) has been lower than at any time prior to 2007 in the satellite-based sea ice record (1979-2012). On a pan-Arctic basis, ice coverage at the September minimum is now about 40% less than in the 1980s, and the ice is younger and thinner than at any time in the period of satellite coverage.

The climate characteristics simulated by climate models have the following key features:

- CMIP3 multi-model mean simulations show similar spatial patterns of increasing mean annual temperature for both the high (A2) and low (B1) emissions scenarios and for three future time periods (2021-2050, 2041-2070, and 2070-2099). Models indicate that these increases are statistically significant across Alaska.

- Mean annual temperature increases are greatest in northwestern Alaska, decreasing southward. The largest simulated warming occurs in the late 21st century under the high (A2) emissions scenario. Differences in the two scenarios (A2 and B1) are especially apparent in winter.

- The range of model-simulated temperature changes is substantial, indicating substantial uncertainty in the magnitude of warming associated with each scenario. However, in each model simulation, the warming is unequivocal and large compared to historical variations. This is also true for all of the derived temperature variables described below.

- Increases of several weeks (15-25 days) are simulated in growing season length for southwestern and south central parts of the state (for the time period 2060-2069 and under both emissions scenarios). In a large portion of southwestern Alaska, the growing season lengthens to more than 200 days, a value currently found only along the southern coastline and subarctic islands.

- An increase in precipitation is simulated to occur across Alaska, with the greatest increases occurring in the northwestern portion of the state. Increases are found to be statistically significant and are greatest in the late 21st century under the high (A2) emissions scenario. However, the range of model-simulated precipitation changes is considerably larger than the multi-model mean change. Thus, there is great uncertainty associated with future precipitation changes in these scenarios.

- Almost all models indicate increases in mean precipitation for all seasons under the high (A2) emissions scenario. For both emissions scenarios, no model indicates a decrease in precipitation in winter. Overall, uncertainty regarding seasonal precipitation changes is great.

- Permafrost degradation is simulated to increase for two future time periods (2040-2049 and 2090-2099) and under both emissions scenarios. Increases cover much of Interior Alaska under the high (A2) emissions scenario. Thawing is more discontinuous in the B1 simulation, although patches of thaw extend as far north as the Brooks Range.

- Climate models simulate large declines in sea ice extent in the Alaskan region. Summer sea ice in the Chukchi Sea disappears between 2030 and 2050 in some models, while winter sea ice in the Bering and Chukchi Seas decrease by more than 50% by the end of the century.

5. REFERENCES

AchutaRao, K., and K.R. Sperber, 2002: Simulation of the El Niño Southern Oscillation: Results from the Coupled Model Intercomparison Project. *Clim. Dyn.,* **19,** 191–209.

ACRC, cited 2012: Temperature Change in Alaska. [Available online at http://climate.gi.alaska.edu/ClimTrends/Change/TempChange.html.]

AMAP, 2011: *Snow, Water, Ice and Permafrost in the Arctic (SWIPA): Climate Change and the Cryosphere.* Arctic Monitoring and Assessment Programme, 538 pp. [Available online at http://amap.no/swipa/CombinedReport.pdf.]

AMSA, 2009: Arctic Marine Shipping Assessment 2009 Report. Arctic Council, 194 pp. [Available online at http://www.pame.is/document-library-single-art/reports-to-saos-and-ministers.]

Arakawa, A., 2004: The cumulus parameterization problem: Past, present, and future. *J. Climate,* **17,** 2493-2525.

Bader D. C., C. Covey, W. J. Gutowski Jr., I. M. Held, K. E. Kunkel, R. L. Miller, R. T. Tokmakian, and M. H. Zhang, 2008: *Climate models: An Assessment of Strengths and Limitations.* U.S. Climate Change Science Program Synthesis and Assessment Product 3.1. Department of Energy, Office of Biological and Environmental Research, 124 pp.

Brohan, P., J. J. Kennedy, I. Harris, S. F. B. Tett, and P. D. Jones, 2006: Uncertainty estimates in regional and global observed temperature changes: A new data set from 1850. *J. Geophys. Res.,* **111,** D12106.

Callaghan, T. V., M. Johansson, O. Anisimov, H. H. Christiansen, A. Instanes, V. Romanovsky and S. Smith, 2011: Changing permafrost and its impacts. *Snow, Water, Ice and Permafrost in the Arctic (SWIPA): Climate Change and the Cryosphere,* Arctic Monitoring and Assessment Programme, 62-85. [Available online at http://amap.no/swipa/CombinedReport.pdf.]

Dufresne, J.-L., and S. Bony. 2008: An assessment of the primary sources of spread of global warming estimates from coupled ocean–atmosphere models. *J. Climate,* **21,** 5135-5144.

Faruch, M. A., H. Hayasaka, and K. Kimura, 2011: Recent anomalous lightning occurrences in Alaska – the case of June 2005. *J. Disaster Res.,* **6,** 321-330.

Hayhoe, K. A., 2010: A standardized framework for evaluating the skill of regional climate downscaling techniques. Ph.D. thesis, University of Illinois, 153 pp. [Available online at https://www.ideals.illinois.edu/handle/2142/16044.]

Hu, F. S., P. E. Higuera, J. E. Walsh, W. L. Chapman, P. A. Duffy, L. B. Brubaker, and M. L. Chipman, 2010: Tundra burning in Alaska: Linkages to climatic change and sea ice retreat. *J. Geophys. Res.,* **115,** G04002.

IPCC, 2000: *Special Report on Emissions Scenarios: A Special Report of Working Group III of the Intergovernmental Panel on Climate Change,* N. Nakicenovic, and R. Swart, Eds., Cambridge University Press, 570 pp.

——, 2007a: *Climate Change 2007: The Physical Science Basis. Contribution of Working Group I to the Fourth Assessment Report of the Intergovernmental Panel on Climate Change,* Solomon, S., D. Qin, M. Manning, Z. Chen, M. Marquis, K.B. Averyt, M. Tignor, and H.L. Miller, Eds., Cambridge University Press, 996 pp.

——, 2007b: *Climate Change 2007: Synthesis Report. Contribution of Working Groups I, II and III to the Fourth Assessment Report of the Intergovernmental Panel on Climate Change,* Pachauri, R. K, and Reisinger, A., Eds., IPCC, 104 pp.

——, cited 2012: IPCC Data Distribution Centre. [Available online at http://www.ipcc-data.org/ddc_co2.html.]

Karl, T. R., G. A. Meehl, C. D. Miller, S. J. Hassol, A. M. Waple, and W. L. Murray, 2008: *Climate Extremes in a Changing Climate, Regions of Focus: North America, Hawaii, Caribbean and US Pacific Islands.* U.S. Climate Change Science Program Synthesis and Assessment Product 3.3. Department of Commerce, NOAA's National Climatic Data Center, 164 pp.

Karl, T. R., J. M. Melillo, T. C. Peterson, and S. J. Hassol, 2009: *Global Climate Change Impacts in the United States.* Cambridge University Press, 188 pp.

Kasischke, E. S., and Coauthors, 2010: Alaska's changing fire regime - implications for the vulnerability of its boreal forests. *Can. J. Forest Res.,* **40,** 1313-1324.

Knutti, R., 2010: The end of model democracy? *Climatic Change,* **102,** 395-404.

Kunkel, K., R. Pielke, and S. Changnon, 1999: Temporal fluctuations in weather and climate extremes that cause economic and human health impacts: A review. *Bull. Am. Meteorol. Soc.,* **80,** 1077-1098.

Larsen, P., S. Goldsmith, O. Smith, M. Wilson, K. Strzepek, P. Chinowsky, and B. Saylor, 2008: Estimating future costs for Alaska public infrastructure at risk from climate change. *Global Environ. Chang.,* **18,** 442-457.

Liston, G. E., and C. A. Hiemstra, 2011: The changing cryosphere: Pan-Arctic snow trends (1979-2009). *J. Climate.,* **24,** 5691-5712.

Medeiros, B., C. Deser, R. A. Tomas and J. E. Kay, 2011: Arctic inversion strength in climate models. *J. Climate,* **24,** 4733-4740.

Meehl, G. A., and Coauthors, 2007: Global climate projections. *Climate Change 2007: The Physical Basis. Contribution of Working Group I to the Fourth Assessment Report of the Intergovernmental Panel on Climate Change,* Solomon, S., D. Qin, M. Manning, Z. Chen, M. Marquis, K.B. Averyt, M. Tignor, and H.L. Miller, Eds., Cambridge University Press, 747-845.

Meier, W. N., J. Stroeve, and F. Fetterer, 2007: Whither Arctic sea ice? A clear signal of decline regionally, seasonally and extending beyond the satellite record. *Ann. Glaciol.,* **46,** 428-434.

Monahan, A.H., and A. Dai, 2004: The spatial and temporal structure of ENSO nonlinearity. *J. Climate,* **17,** 3026–3036.

NOAA, cited 2012: Cooperative Observer Program. [Available online at http://www.nws.noaa.gov/om/coop/.]

Norton, C. W., P.-S. Chu, and T. A. Schroeder, 2011: Projecting changes in future heavy rainfall events for Oahu, Hawaii: A statistical downscaling approach. *J. Geophys. Res.,* **116,** D17110.

Overland, J. E., and M. Wang, 2007: Future regional Arctic sea ice declines. *Geophys. Res. Lett.,* **34,** L17705.

Overland, J. E., M. Wang, N. A. Bond, J. E. Walsh, V. M. Kattsov, and W. L. Chapman, 2011: Considerations in the selection of global climate models for regional climate projections: The Arctic as a case study. *J. Climate,* **24,** 1583-1597.

PCMDI, cited 2012: CMIP3 Climate Model Documentation, References, and Links. [Available online at http://www-pcmdi.llnl.gov/ipcc/model_documentation/ipcc_model_documentation.php.]

Perovich, D. K., and J. A. Richter-Menge, 2009: Loss of Sea Ice in the Arctic. *Annu. Rev. Mar. Sci.*, **1,** 417-441.

Randall, D.A., and Coauthors, 2007: Climate models and their evaluation. *Climate Change 2007: The Physical Basis. Contribution of Working Group I to the Fourth Assessment Report of the Intergovernmental Panel on Climate Change*, Solomon, S., D. Qin, M. Manning, Z. Chen, M. Marquis, K.B. Averyt, M. Tignor, and H.L. Miller, Eds., Cambridge University Press, 590-662.

Shulski, M., and G. Wendler, 2007: *The Climate of Alaska.* University of Alaska Press, 208 pp.

SNAP, cited 2012: Scenarios Network for Alaska and Arctic Planning. [Available online at http://www.snap.uaf.edu/.]

Smith, S., V. Romanovsky, A. Lewkowicz, C. Burn, M. Allard, G. Clow, K. Yoshikawa and J. Throop, 2010: Thermal state of permafrost in North America: A contribution to the international polar year. *Permafrost and Periglac. Process.*, **21,** 117-135.

Stewart, B. C., 2011: Changes in frequency of extreme temperature and precipitation events in Alaska. M Master's thesis, University of Illinois, 65 pp. [Available online at https://www.ideals.illinois.edu/handle/2142/24093.]

Stroeve, J. C., M. C. Serreze, M. M. Holland, J. E. Kay, J. Malanik, and A. P. Barrett, 2011: The Arctic's rapidly shrinking sea ice cover: a research synthesis. *Climatic Change*, **110,** 1005-1027.

Tebaldi, C., J. M. Arblaster, and R. Knutti, 2011: Mapping model agreement on future climate projections. *Geophys. Res. Lett.*, **38,** L23701.

UIUC, cited 2012: The Cryosphere Today. [Available online at http://arctic.atmos.uiuc.edu/cryosphere/.]

Walsh, J. E., W. L. Chapman, V. Romanovsky, J. H. Christensen, and M. Stendel, 2008: Global climate model performance over Alaska and Greenland. *J. Climate*, **21,** 6156-6174.

Wilby, R. L., and T. Wigley, 1997: Downscaling general circulation model output: a review of methods and limitations. *Prog. Phys. Geog.*, **21,** 530.

WRCC, cited 2011: Climate of Alaska. [Available online at http://www.wrcc.dri.edu/narratives/ALASKA.htm.]

6. ACKNOWLEDGEMENTS

We acknowledge the modeling groups, the Program for Climate Model Diagnosis and Intercomparison (PCMDI) and the WCRP's Working Group on Coupled Modelling (WGCM) for their roles in making available the WCRP CMIP3 multi-model dataset. Support of this dataset is provided by the Office of Science, U.S. Department of Energy. Analysis of the CMIP3 GCM simulations was provided by Michael Wehner of the Lawrence Berkeley National Laboratory and by Jay Hnilo of the Cooperative Institute for Climate and Satellites (CICS). Document support was provided by Fred Burnett and Clark Lind of TBG Inc. Additional programming and graphical support was provided by Scott Stevens and Andrew Buddenberg of CICS, Greg Dobson of the University of North Carolina-Asheville, and Byron Gleason of NOAA's National Climatic Data Center (NCDC). The downscaling of the climate models was performed by Tom Kurkowski, Michael Lindgren, Matthew Leonawicz, William Chapman, Tracy Rogers and other members of the Scenarios Network for Alaska and Arctic Planning, University of Alaska, Fairbanks.

A partial listing of reports appears below:

NESDIS 102 NOAA Operational Sounding Products From Advanced-TOVS Polar Orbiting Environmental Satellites. Anthony L. Reale, August 2001.

NESDIS 103 GOES-11 Imager and Sounder Radiance and Product Validations for the GOES-11 Science Test. Jaime M. Daniels and Timothy J. Schmit, August 2001.

NESDIS 104 Summary of the NOAA/NESDIS Workshop on Development of a Coordinated Coral Reef Research and Monitoring Program. Jill E. Meyer and H. Lee Dantzler, August 2001.

NESDIS 105 Validation of SSM/I and AMSU Derived Tropical Rainfall Potential (TRaP) During the 2001 Atlantic Hurricane Season. Ralph Ferraro, Paul Pellegrino, Sheldon Kusselson, Michael Turk, and Stan Kidder, August 2002.

NESDIS 106 Calibration of the Advanced Microwave Sounding Unit-A Radiometers for NOAA-N and NOAA-N=. Tsan Mo, September 2002.

NESDIS 107 NOAA Operational Sounding Products for Advanced-TOVS: 2002. Anthony L. Reale, Michael W. Chalfant, Americo S. Allegrino, Franklin H. Tilley, Michael P. Ferguson, and Michael E. Pettey, December 2002.

NESDIS 108 Analytic Formulas for the Aliasing of Sea Level Sampled by a Single Exact-Repeat Altimetric Satellite or a Coordinated Constellation of Satellites. Chang-Kou Tai, November 2002.

NESDIS 109 Description of the System to Nowcast Salinity, Temperature and Sea nettle (*Chrysaora quinquecirrha*) Presence in Chesapeake Bay Using the Curvilinear Hydrodynamics in 3-Dimensions (CH3D) Model. Zhen Li, Thomas F. Gross, and Christopher W. Brown, December 2002.

NESDIS 110 An Algorithm for Correction of Navigation Errors in AMSU-A Data. Seiichiro Kigawa and Michael P. Weinreb, December 2002.

NESDIS 111 An Algorithm for Correction of Lunar Contamination in AMSU-A Data. Seiichiro Kigawa and Tsan Mo, December 2002.

NESDIS 112 Sampling Errors of the Global Mean Sea Level Derived from Topex/Poseidon Altimetry. Chang-Kou Tai and Carl Wagner, December 2002.

NESDIS 113 Proceedings of the International GODAR Review Meeting: Abstracts. Sponsors: Intergovernmental Oceanographic Commission, U.S. National Oceanic and Atmospheric Administration, and the European Community, May 2003.

NESDIS 114 Satellite Rainfall Estimation Over South America: Evaluation of Two Major Events. Daniel A. Vila, Roderick A. Scofield, Robert J. Kuligowski, and J. Clay Davenport, May 2003.

NESDIS 115 Imager and Sounder Radiance and Product Validations for the GOES-12 Science Test. Donald W. Hillger, Timothy J. Schmit, and Jamie M. Daniels, September 2003.

NESDIS 116 Microwave Humidity Sounder Calibration Algorithm. Tsan Mo and Kenneth Jarva, October 2004.

NESDIS 117 Building Profile Plankton Databases for Climate and EcoSystem Research. Sydney Levitus, Satoshi Sato, Catherine Maillard, Nick Mikhailov, Pat Cadwell, Harry Dooley, June 2005.

NESDIS 118 Simultaneous Nadir Overpasses for NOAA-6 to NOAA-17 Satellites from 1980 and 2003 for the Intersatellite Calibration of Radiometers. Changyong Cao, Pubu Ciren, August 2005.

NESDIS 119 Calibration and Validation of NOAA 18 Instruments. Fuzhong Weng and Tsan Mo, December 2005.

NESDIS 120 The NOAA/NESDIS/ORA Windsat Calibration/Validation Collocation Database. Laurence Connor, February 2006.

NESDIS 121 Calibration of the Advanced Microwave Sounding Unit-A Radiometer for METOP-A. Tsan Mo, August 2006.

NESDIS 122   JCSDA Community Radiative Transfer Model (CRTM). Yong Han, Paul van Delst, Quanhua Liu, Fuzhong Weng, Banghua Yan, Russ Treadon, and John Derber, December 2005.

NESDIS 123   Comparing Two Sets of Noisy Measurements. Lawrence E. Flynn, April 2007.

NESDIS 124   Calibration of the Advanced Microwave Sounding Unit-A for NOAA-N'. Tsan Mo, September 2007.

NESDIS 125   The GOES-13 Science Test: Imager and Sounder Radiance and Product Validations. Donald W. Hillger, Timothy J. Schmit, September 2007.

NESDIS 126   A QA/QC Manual of the Cooperative Summary of the Day Processing System. William E. Angel, January 2008.

NESDIS 127   The Easter Freeze of April 2007: A Climatological Perspective and Assessment of Impacts and Services. Ray Wolf, Jay Lawrimore, April 2008.

NESDIS 128   Influence of the ozone and water vapor on the GOES Aerosol and Smoke Product (GASP) retrieval. Hai Zhang, Raymond Hoff, Kevin McCann, Pubu Ciren, Shobha Kondragunta, and Ana Prados, May 2008.

NESDIS 129   Calibration and Validation of NOAA-19 Instruments. Tsan Mo and Fuzhong Weng, editors, July 2009.

NESDIS 130   Calibration of the Advanced Microwave Sounding Unit-A Radiometer for METOP-B. Tsan Mo, August 2010.

NESDIS 131   The GOES-14 Science Test: Imager and Sounder Radiance and Product Validations. Donald W. Hillger and Timothy J. Schmit, August 2010.

NESDIS 132   Assessing Errors in Altimetric and Other Bathymetry Grids. Karen M. Marks and Walter H.F. Smith, January 2011.

NESDIS 133   The NOAA/NESDIS Near Real Time CrIS Channel Selection for Data Assimilation and Retrieval Purposes. Antonia Gambacorta, Chris Barnet, Walter Wolf, Thomas King, Eric Maddy, Murty Divakarla, Mitch Goldberg, April 2011.

NESDIS 134   Report from the Workshop on Continuity of Earth Radiation Budget (CERB) Observations: Post-CERES Requirements. John J. Bates and Xuepeng Zhao, May 2011.

NESDIS 135   Averaging along-track altimeter data between crossover points onto the midpoint gird: Analytic formulas to describe the resolution and aliasing of the filtered results. Chang-Kou Tai, August 2011.

NESDIS 136   Separating the Standing and Net Traveling Spectral Components in the Zonal-Wavenumber and Frequency Spectra to Better Describe Propagating Features in Satellite Altimetry. Chang-Kou Tai, August 2011.

NESDIS 137   Water Vapor Eye Temperature vs. Tropical Cyclone Intensity. Roger B. Weldon, August 2011.

NESDIS 138   Changes in Tropical Cyclone Behavior Related to Changes in the Upper Air Environment. Roger B. Weldon, August 2011.

NESDIS 139   Computing Applications for Satellite Temperature Datasets: A Performance Evaluation of Graphics Processing Units. Timothy F.R. Burgess and Scott F. Heron, December 2011.

NESDIS 140   Microburst Nowcasting Applications of GOES. Kenneth L. Pryor, September 2011.

NESDIS 141   The GOES-15 Science Test: Imager and Sounder Radiance and Product Validations. Donald W. Hillger and Timothy J. Schmit, November 2011.

NOAA SCIENTIFIC AND TECHNICAL PUBLICATIONS

The National Oceanic and Atmospheric Administration was established as part of the Department of Commerce on October 3, 1970. The mission responsibilities of NOAA are to assess the socioeconomic impact of natural and technological changes in the environment and to monitor and predict the state of the solid Earth, the oceans and their living resources, the atmosphere, and the space environment of the Earth.

The major components of NOAA regularly produce various types of scientific and technical information in the following types of publications

PROFESSIONAL PAPERS – Important definitive research results, major techniques, and special investigations.

CONTRACT AND GRANT REPORTS – Reports prepared by contractors or grantees under NOAA sponsorship.

ATLAS – Presentation of analyzed data generally in the form of maps showing distribution of rainfall, chemical and physical conditions of oceans and atmosphere, distribution of fishes and marine mammals, ionospheric conditions, etc.

TECHNICAL SERVICE PUBLICATIONS – Reports containing data, observations, instructions, etc. A partial listing includes data serials; prediction and outlook periodicals; technical manuals, training papers, planning reports, and information serials; and miscellaneous technical publications.

TECHNICAL REPORTS – Journal quality with extensive details, mathematical developments, or data listings.

TECHNICAL MEMORANDUMS – Reports of preliminary, partial, or negative research or technology results, interim instructions, and the like.

U.S. DEPARTMENT OF COMMERCE
National Oceanic and Atmospheric Administration
National Environmental Satellite, Data, and Information Service
Washington, D.C. 20233